# The Story of the Hoover Dam

Hoover Dam, among the greatest of modern engineering projects, was so gigantic in its dimensions and so vast in its economic significance that progress of its construction during 1931-1935 was followed with interest throughout the world. In keeping with its long-standing editorial policy, *Compressed Air Magazine* during those years issued a series of five booklets giving its readers a progressive history of the undertaking.

The material from those booklets is reproduced here except for the advertising and material about the Colorado River Aqueduct. The early history of the Colorado River is recounted, along with the developments leading up to the selection of the Hoover Dam site. The founding of Boulder City is followed by a description of the construction of the four diversion tunnels which carried the Colorado River water through the walls of Black Canyon during dam construction.

Also described is the extensive railroad system, the gravel-treating and concrete-mixing plants, the cofferdams, the aerial cableways, the penstocks, and how the concrete was poured and cooled. The human side is not neglected; these articles tell how government engineers and surveyors and hundreds of laborers conducted exacting, perilous work.

As the last of the booklets went to press, Hoover Dam was substantially finished. Six Companies, Inc. had completed their $49 million contract, and the government had accepted their work. Lake Mead began to form behind the dam, and the threat of flood downstream was permanently gone. The metropolitan aqueduct would soon supply water to thirteen southern California cities and the power transmission line to Los Angeles was nearly ready. The once unruly Colorado River would be effectively brought under man's control.

Nevada Publications
Box 15444
Las Vegas, Nevada 89114


Printed in the United States of America. ISBN 913814-79-2

High above the Colorado River these workmen are placing anchorages from which drillers will swing by ropes in scaling down the walls of Black Canyon.

# THE STORY of the HOOVER DAM

## Table of Contents

Drawing by F. W. Egloffstern from sketch by Lieut. Ives

*Black Canyon as it seemed to Lieut. Joseph C. Ives in 1857.*

Drawing by J. J. Young from sketch by H. B. Mollhausen
**Ives' little steamer "Explorer" making her way through Mohave Canyon.**

# America's Wonder River---The Colorado

## *How Nature Formed This Amazing Watercourse and How Man Has Had to Face Many Hazards in Discovering It and in Completing Its Exploration*

**By R. G. SKERRETT**

NEARLY four hundred years ago, a white man gazed for the first time upon the Colorado River. No one will ever know who was the first human being to look upon that stream.

That amazing waterway might have remained unknown to white men for centuries longer had not an insatiable greed for gold spurred the *conquistadores* to seek still greater riches than they had been able to wring from the hapless Montezuma.

Since Hernando de Alarcon discovered the Colorado in 1540, and struggled valiantly a considerable distance upstream against its muddy and turbulent currents, very little has been done to control its flow or to convert its tremendous forces into a form of energy that could be put to commensurate and befitting services.

Now, by reason of collaboration on the part of the seven states directly affected and the Federal Government, the construction of monumental Hoover Dam is underway in Black Canyon. If all goes well, that titanic structure will be finished five years hence. Thus nearly 400 years after its discovery, the rampageous Colorado will be held under control by an engineering checkrein for conservation, for irrigation, and for the generation of many thousands of horsepower of electrical energy that will be used throughout a wide radius to do a vast deal of work.

The real significance of this approaching transformation can be better understood if we delve a bit into the dim past of this wonderful waterway. Only by so doing can we grasp why the river is as it is today—unique not only because of its magnificent and even appalling scenery but also because it is unrivaled as an impressive example of what the erosive forces of wind and water can do when acting over a long, long stretch of time. The Colorado has been millions of years in the making.

Drawing by F. S. Dellenbaugh
**How Alarcon's caravels battled with the bore at the mouth of the Colorado in 1540.**

### *The Colorado and Its Course*

From source to mouth, the Colorado has a total length of 1,700 miles; and today it drains an area of approximately 244,000 square miles. Once upon a time the Colorado had a much shorter route to follow before it reached the sea; and then, as now, something like 75 per cent of the water discharged by it had its origin in the glacial and snow-capped mountain ranges forming the watershed of its northern basin. Much of the hundreds of miles of marvelous canyon sections of the present then lay submerged beneath a comparatively shallow sea—an inland extension of what we now know as the Gulf of California. As the ages passed, the river carried to its estuary enormous quantities of sand

Drawing by J. J. Young from sketch by H. B. Mollhausen

**Steamer "Explorer", commanded by Lieut. Ives, making her way up the Colorado in 1857. Chimney Peak is seen in the distance.**

gathered by its ramifying tributaries, and thus layer by layer a blanket of ever increasing depth was laid on the water bed. Then came a period of momentous change.

Internal pressures of irresistible magnitude started an uplift that continued until the Sierras reared their shoulders high above the sea, and at the same time was raised the estuary into which what we call the Colorado then emptied—transforming that water bed into dry land. Then and there the river began to cut its way through the sand to the sea beyond—thus reversing the process by which it had previously formed that deposit, thousands of feet in thickness.

Where the course of the river was a leisurely one, it carved for itself a generously broad right of way; and where its descent was abrupt and its flow rapid, the stream eroded a narrow and correspondingly deeper channel for itself. Geologists tell us that the region was subjected to three great uplifts, each uplift being from 2,000 to 3,000 feet above the sea. There are evidences that the region was successively raised and submerged. These tremendous alterations occurred over the inconceivably protracted period of five geological eras—the Colorado, the while, tirelessly cutting a course through sands or rocky structures as they arose athwart its path. In this manner the canyons have been formed. Today, the river is eroding a channel through the great basic stratum of hard black gneiss. The Grand Canyon, in particular, bears mute and appalling testimony to the ages and ages that have gone their several ways into the dim past since the river began its tireless and irresistible work. Following the Grand Canyon inward from its outermost rim, the flat-topped buttes are like so many gigantic steps that indicate the successive strata of rock that the river has cut away before digging that vast scar or gorge through which the stream traces its present course deep below its innermost crest. One must not forget that the work of the river has been supplemented by the erosive sweep of winds and infrequent rains—these agencies still continuing to modify the modeling of the buttes exposed to them.

The magnitude of this total erosive action is incomprehensibly great. We are authoritatively informed that no part of the whole region has been worn down, degraded, less than 1,000 feet; and there is one area of more than 200,000 square miles in extent that has been degraded on an average of more than 6,000 feet. This erosion has been wrought in various kinds of limestone, sandstone, quartzite, and granite—not to mention a stretch of 50 miles where the river has cut its course through lava poured into the gorge to a depth of 200 or 300 feet by erstwhile active volcanoes.

Softly falling snowflakes, clouds spilling their raindrops at lower levels have, together, brought about this staggeringly wonderful metamorphosis. The melting snows, the pattering or beating rains have changed rocky surfaces to sands, and mountain rills have carried the sands into neighboring creeks. The creeks, in their turn, have conveyed the sands to rivers, and the rivers, in their turn, have borne the sands to the Colorado. At each step the abrasive action of the sand has been intensified. In short, the river, after its manner, has performed in a magnified degree not unlike the relatively diminutive saws that man employs in the cutting of blocks of granite or stone of lesser hardness. And the sand that has been used by the Colorado in carving its canyons has finally come to rest in the far-flung delta that has been built up contiguous to the Gulf of California. This burden of silt averages annually sufficient to cover an area of 80,000 acres to a depth of one foot; and this rate of deposition has been going on for an untold period of time. In other words, the very substance of the great plateau through which the Colorado has made its way is being moved ceaselessly from the distant interior to the border of the Pacific Ocean.

The Grand Canyon of the Colorado lies between the Paria River on the north and Grand Wash Cliffs, 278 miles downstream to the south and

U. S. Bureau of Reclamation

**All that remains of the old Town of Callville built by the Mormons in 1864. Callville was the head of navigation for craft ascending the river from Yuma.**

G. H. Bishop

**Spectacular section of Marble Gorge illustrating the vast and continuous erosive action of the Colorado River.**

U. S. Geological Survey and E. L. Kolb

**Left and right—Sections of Marble Gorge. Center—Engineers of the United States Geological Survey at work in Marble Gorge.**

west. Its greatest width is thirteen miles, and at its deepest point it measures 6,000 feet from rim to water level. Where the various canyons penetrate to the supporting granitic formation, they reveal the very adolescent days of Mother Earth. The walls present a perfect picture of much of the continent's geological past; and the picture is awesome by reason of the story it tells and because of the magnificent scale upon which it is portrayed. In the presence of such a record, man and his measures of time dwindle into the infinitesimal. One stands speechless before the work of the tireless Colorado.

### *Finding and Exploring the River*

Cortez filled the strong boxes of Spain with treasure stripped from the Aztecs; but his aggressiveness and his success multiplied his enemies in Spain as well as in America. To win royal favor he knew that he would have to garner added wealth for his king and for himself. But how to do so was a problem. Then, when the outlook seemed darkest, Alva Nunez Cabeza de Vaco returned from his wanderings between the Mississippi and the headwaters of the Rio Grande. De Vaca repeated the tales of a friendly Amerind or Indian who told of rich cities, visited by him when a boy, that lay somewhere to the westward oi the Rio Grande. The Amerind had seen silversmiths fashioning jewelry and plate of precious metals in large quantities. These tales were amplified by an imaginative Negro, one Estavan, also a member of de Vaca's party. This was in 1536; and instantly Cortez was moved to action—he wanted to be the first to reach and to plunder the so-called Seven Cities of Cibola.

With funds obtained by pawning his wife's jewels, Cortez equipped three caravels commanded by Francisco de Ulloa; and dispatched them from Acapulco in quest of a point northward that would make it easy for the Spaniards to reach the reputedly opulent cities from the coast. Nothing untoward happened until Ulloa found his craft in the midst of rapidly shoaling water and swift currents and with shores closing in on them where the open sea was expected. Ignorant of the coast, Ulloa had reached the headwaters of the Gulf of California—the existence of which was unsuspected. Grace, indeed, was the peril when the incoming tide, in the form of a high and roaring wave, swept toward the vessels—that startling phenomenon being due to a range of 36 feet between high and low water and to the narrowed passage the water had to follow during the changing tides.

As Ulloa wrote in his tragic report to Cortez: "We always found more shallow water and the sea thick, black, and very muddy. . . . We rode all night in five fathom of water, and we perceived the sea to run with so great a rage into the land that it was a thing much to be marvelled at; and with the like fury it returned back again with the ebb, during which time we found eleven fathom water, and the flood and ebb continued from five to six hours." Whence came the waters that poured seaward with the ebb was a matter of wondering speculation; but neither Ulloa nor any of his followers realized that they were at the mouth of a great river.

Ulloa was no coward; but he was an experienced seaman who recognized his perilous predicament. Therefore, instead of waiting for his caravels to be battered to bits by a recurrence of the bore, he sailed southward to more open water. From there he sent one of his craft back to Acapulco with the story of his failure, and with this done he rounded the southern tip of Baja California and headed away with his remaining ships on a voyage that carried all hands to eternity.

Ulloa's failure, and the subsequent killing of the Negro Estavan when he and Friar Marcos tried to reach the Seven Cities of Cibola by an overland route, did not deter the Spaniards—it rather added fuel to the flame of conquest that burned in their breasts. Accordingly, Cortez' viceregal successor, Mendoza, equipped two expeditonary forces in 1540 for the dual purpose of exploration and conquest. Mendoza believed he could in this way make certain of success. Francisco Vasquez de Coronado lead the land force that was to follow the coast northward from Compostela; and the nautical force of three ships commanded by Hernando de Alarcon sailed from Acapulco and reached the head of the Gulf of California towards the end of August. Coronado arrived at Cibola in July, and instead of a place of riches he found a community of adobe huts much like those common to the region today. Two years later Coronado returned to Mexico virtually empty handed and without having effected contact with Alarcon; but the latter achieved the discovery of the Colorado River and blazed the way for subsequent adventures that only men of great courage and determination could have carried to their several conclusions.

Alarcon was confronted with the same harassing and hazardous conditions that had halted Ulloa; and, had he heeded the advice of his pilots, he, too, would have turned back rather than to risk the destruction of his ships. But, because Mendoza had ordered him to discover the secret of the gulf, Alarcon resolved to push onward despite the risks involved. While the bore swept Alarcon's boats about like toys, and nearly turned them over again and again, still all his caravels survived virtually undamaged; and, when the tide served, the vessels were maneuvered into the channel. As Alarcon reported: "It pleased God that after this sort we came to the very bottom of the bay, where we found a very mighty river, which ran with so great a fury of a stream, that we could hardly sail against it."

When he had his craft where they could be moored and protected against the direct force of the bore, Alarcon provisioned and armed two boats for a trip farther up the river. The current soon proving too strong to be rowed against, most of the crews were landed and set to the task of towing them. It was hard work for men unfamiliar with that form of labor: and when inquisitive and seemingly hostile Indians gathered on the banks, Alarcon had the wit to make friendly gestures instead of firing his guns to alarm them. Furthermore, by suitable pantomime, he made them understand that the Spaniards represented the Indian's sovereign lord, the Sun. From that time on for a number of days the natives vied with one another to put their shoulders to the tow lines. Alarcon made two trips up the waterway, and the highest point reached by him on his second expedition was somewhere about 50 miles above the mouth of the Gila River. Failing then to make contact with any of Coronado's lieutenants, Alarcon returned to his ships, which were badly damaged by the *teredo*, and hastened southward to a suitable port for repairs.

Of Coronado's force only two men reached the Colorado. One of his officers, Melchior Diaz, came to the river after Alarcon had departed; and to Don Lopez de Cardenas, a rollicking member of that expedition, belongs the credit of being the first white man to look down into the depths of the Grand Canyon. He probably did this, so it is believed by competent authorities, at a point somewhere between Diamond Creek and Kanab Canyon. Thus far we have dealt with the actual discovery of the Colorado, so named because of the reddish color of its waters. Now let us touch somewhat briefly upon subsequent explorations over succeeding centuries that finally rounded out our knowledge of the Colorado so recently as 1924.

Following in the footsteps of the gold-seeking *conquistadores* came those gentle-hearted *padres* whose primary purpose was the saving of souls. It was Father Garces who first made continual use of the name Colorado in describing the river. To Fathers Garces and Escalante we are indebted for accounts of different parts of it; and the latter, after traversing a desperately difficult canyon section of that waterway, discovered a place where it could be forded. The route now known as the Crossing of the Fathers bears testimony to the good *padres'* undaunted persistence in the face of appalling odds. This was back in 1776. Decades passed with no records of white explorers along the Colorado until James O. Pattie, a venturesome trapper, followed the river all the way from its mouth to its headwaters—being in all probability the first white man, if not the first person, to do so. Pattie's journey was made during the "twenties" of the past century. Pattie traveled both on foot and mounted.

To Gen. William Henry Ashley, a fur trader, belongs the credit of being the first to go by boat down any of the canyons of the Colorado. Most of his perilous voyage was down the Green River, and was made in 1825 in craft covered with buffalo skins instead of wooden sheathing. The account of this trip abounds in thrills. Several members of his party were drowned; and the expedition traversed some

U. S. Geological Survey

**Top—Where boats of the Survey journeyed overland at a point in Soap Creek Rapids. Center—Running the rapids below Havasu Creek. Bottom—Repairing one of the boats after reaching Badger Creek.**

of the worst rapids in the canyons of the Colorado system. Of the excellent work of Lieut. W. H. Hardy, R. N., we can only refer to the fact that he explored the Gulf of California in 1825 for a pearl-and-coral fishery association. To him the world is indebted for the first authentic plan of the lowermost part of the Colorado River as well as of the Gila for a few miles above its confluence with the Colorado. Lieutenant Hardy's 25-ton schooner had to battle with the bore as had his Spanish predecessors; and by good fortune and expert maneuvering he managed to save his vessel from destruction.

Without any intent to belittle their performance, we must touch only briefly upon the trip down the canyons of the Green River made by William Manly and six companions —all "forty-niner" teamsters who abandoned bull-whacking to venture afloat upon the "River of Mystery". They were intent upon reaching the Pacific Coast by what they conceived to be a fortuitous and shorter route than that followed by prairie schooners. The craft used was little better than a scow. All went well, despite some narrow squeaks at several points, until the scow was smashed for keeps on the rocks in the rapids of Ashley Falls. From there on those high-spirited young men made their way overland.

The continual lure of gold in California, and the need of a military post near the junction of the Gila and the Colorado, caused the United States Government to make its first attempt to explore the Colorado in 1850. Lieut. George H. Derby of the Army was ordered to enter the Colorado as far as the Gila in a sailing vessel of 120 tons. The *Invincible*, as she was named was handled by a veritable sailor who, when the roaring flood tide hit her and she seemed doomed, saved the craft by prompt and decisive action. Derby's penetration to the Gila led shortly afterwards to commercial voyaging upon the lower reaches of the Colorado. George A. Johnson, the pioneer in this field, built a flotilla of barges and later operated one or more stern-wheel steamers on the river above Yuma. Johnson's work undoubtedly inspired the Government expedition of 1857, which was commanded by Lieut. Joseph C. Ives of the Army.

Ives' vessel, the *Explorer*, was delivered dismantled at the head of the Gulf of California; and there, in the muddy delta of the river, a basin was excavated in the soft ground and the assembling of the stern-wheeler was begun. In about a month the craft was so far finished that steam could be raised in her boiler and her engine turned over. Five days later, at high tide, with a passage cleared from the basin to the waterway, the *Explorer* was successfully launched. In the course of the work cut out for him, Ives navigated his vessel upstream as far as the mouth of Black Canyon. There a sunken rock blocked the way and just missed damaging her irreparably. That ended further upstream exploration with the steamer. When suitably repaired, the return journey was begun. The *Explorer* was disposed of at Yuma; and it is interesting to point out that her hulk was recently uncovered in the sands of the delta a number of miles from the present course of the river—the erstwhile channel having been shifted during the flood period of 1909.

Most of what we now know about the Colorado is mainly due to the daring of those explorers who successively ventured upon the river between 1869 and 1924. Some of them faced, with other perils, death by starvation; and in more than one instance they headed with grim determination into the unknown—an unknown reputed at points to entail plunges into the very bowels of the earth. Our geological, hydrographic, and photographic understanding of the Colorado, may, without disparagement to others, be attributed to the discoveries and the activities of Maj. J. W. Powell, Lieut. George N. Wheeler, Frank M. Brown, Robert Brewster Stanton, Nathan Galloway, Julius F. Stone, the Kolb brothers, and E. C. LaRue, with his associates of the United States Geological Survey. LaRue's explorations were numerous and comprehensive, and covered the period between 1914 and 1924.

The work of certain of the foregoing explorers was epochal in its revelations, and members of some of the parties lost their lives in their splendid efforts to strip the Colorado of the very last of its mysteries. As we shall see in due season, all this courageous venturing was more or less necessary before any successful steps could be taken to curb the Colorado, which has been running its unbridled course for ages. Even now, one wonders at the confidence of Lilliputian man to bind this aquatic Gulliver.

All Photos U. S. Bureau of Reclamation

**Sharp's Heading, where water is drawn from the Colorado to irrigate farmlands in the Imperial Valley. Inserts—What irrigation has done to make arid acres abundantly productive.**

# America's Wonder River---The Colorado

## *Something About Its Continually Changing Delta and the Adjacent Arid Region Which It Has Created and Which Man Is Now Making Abundantly Fruitful*

**By R. G. SKERRETT**

THE Colorado is an anomaly among the great rivers of the world. In addition to certain unique physical features, it has differed radically from other waterways of comparable size because its main stream has generally repelled man's efforts to put it to his use.

The majority of large rivers have lent themselves to utilization in one way or another. They have faciliated exploration; they have aided man in many subsequent practical ways; and, finally they have stimulated industry and commerce and the development of the contiguous watershed. In well-nigh all these respects the Colorado has stood forth a conspicuous contrast; and after the hundreds of years that the white man has known about the river it still is of service to him in only a small measure of its potential capacity. These facts emphasize the problem now squarely faced in building the Hoover Dam so that the stream can be controlled and utilized in ways that so long seemed impossible.

### *What the Aborigines Did*

Cliff dwellers and other early Indian occupants of the territory, drained by some of the Colorado's tributaries, put those rivers to use in irrigating contiguous lands. Those aborigines set the pace in this particular a thousand or more years ago; and in many instances the ditches they dug have survived to the present day. Those Indians were wise: they did not try to tap the main body of the Colorado. They know from many appalling examples just what the river could do in flood periods when its waters were on their annual rampage. Neither did they venture down into the canyon region with the thought of cultivating and irrigating any arable areas that might be found close to the stream.

On the other hand, the Cocopahs and the Yumas early began the growing of crops on the delta lands adjacent to the mouth of the Colorado; but in doing so they took conditions as they found them from season to season and made no effort to change the ways of the river. Where and when the delta lands in due time emerged from the receding floods, the Indians cast their seed, without tilling, upon the wet mud. The method, of course, was primitive like the people, but even so the Indians were able to raise substantial crops of beans and maize and squashes by reason of the fertility of the saturated silt and the wealth of sunshine which shone upon the region. Those aborigines did not know it, but they were doing exactly what the *fellahin* of Egypt had been doing since the days of the Pharaohs each year as the flood waters of the Nile went their way seaward and left vast sodden areas exposed for planting. It might here be mentioned that the delta of the Colorado and the delta of the Nile are much alike climatically—both are subtropical.

The comparison between the two rivers is a limited one. The rise and the fall of the waters of the Nile, as recorded for nearly 5,000 years, is remarkably uniform both in

**Point on the west bank of the Colorado where the third heading, cut by the California Development Company, caused a break that was closed only after months of desperate efforts.**

degree and in time of occurrence. The Colorado, however, varies widely in the volume of water carried by it at the period of maximum flow; and the time of its highest water cannot be foretold with certainty. In short, the river is in a continual state of unstable equilibrium. In the past this characteristic made farming for the Indians a matter of considerable speculation both as to when they could sow their seed and where it would be possible for them to do so. The Indians at least proved conclusively that the delta lands of the Colorado were extremely fertile.

Before we tell how the modern farmer has undertaken to develop lands formed by the Colorado more or less in the neighborhood of its present mouth, let us see how those productive lands came into being and long constituted a veritable desert region until an imaginative white man envisioned their potential agricultural value. The story of their creation is one more evidence of the unstable nature of the Colorado River.

The Colorado carries toward the sea each year an enormous amount of solid matter in suspension; and measurements made with much care over a span of years have revealed that the silt content of the stream at Yuma amounts every twelvemonth to substantially 105,000 acre-feet. That is to say, the average silt content is equal to 0.62 per cent by volume; and this silt, when compacted, has a volume of 170,000,000 cubic yards! This output has been issuing from the mouth of the Colorado for ages, and has repeatedly transformed the topography of the delta: the river has made and unmade the channels by which its flow has reached the open waters of the Gulf of California. This is understandable, because the silt-formed land or river bed offers only a trifling opposition to the erosive action of the stream—a thin thread of water soon becoming a sizable one and, in its turn, growing rapidly to the proportions of a rampant river.

The Colorado, when carrying its greatest burden of solid matter, may be likened to a tremendous hydraulic dredge engaged in drawing up silt at one point to distribute it to other points for the upbuilding of land. There have been times in the history of the river when the alluvial flow was exceptionally large; and on those occasions the stream wrought the most pronounced changes in its delta. Remember that the average yearly discharge of silt is equivalent in volume to all the earth removed in digging the Panama Canal! At some period in its history—prior to the traditions of the Indians in that part of the continent, the Colorado probably emptied directly into the Gulf of California not far below Yuma of today. When that was the case, the Gulf of California extended nearly 150 miles north of its present northernmost limits, and the Colorado discharged into the gulf something like 100 miles south of the head of that body of water.

**Southern Pacific Railway bridge at Yuma, in 1916, when the flood was at its height and had a flow of 240,000 second-feet.**

### *Genesis of Imperial Valley*

Year by year, the Colorado built up a fan-shaped delta that widened and lengthened as the river spread out its burden of silt upon the bed of the gulf. Eventually—probably initiated by some season of unusually heavy alluvial outpouring, it started the growth of a jetty of mud on

**Striking evidence of the havoc wrought by the Colorado River when inundating Imperial Valley. The banks cut by the water range from 30 to 40 feet in height.**

its western flank and southward from its mouth. Gradually this jetty grew until it became in effect a broad, low dike across the gulf, and thus cut off and isolated the area of which the Imperial Valley forms a part. In the course of centuries the impounded water evaporated, until only a remnant of the erstwhile volume remained in the deepest section commonly known as the Salton Sink or Salton Sea. The whole isolated area slopes from the river toward the sink, which lies 278 feet below the normal surface of the Pacific.

While building up the barrier just mentioned the river, incidentally, elevated its own bed by successive deposits of silt. Today, the Colorado flows through a delta which, in some sections, is 100 feet above the sea. During flood periods, the river may wander in one direction or another by reason of this elevation and the character and conformation of the delta. In brief, it is likely now as it has been in the past to spill over its banks anywhere during seasonal high water. When the river does this, it piles mud upon its banks and tends to raise the entire flooded area by new deposits of silt.

Conversely, as the flood waters diminish and drop, the Colorado commonly returns to the channels just previously used by it—the receding waters closing the outlets made during the flood by deposits of mud. On the other hand, the river may continue to follow a new channel cut during the high-water period; and numerous ancient channels give visible evidence of how diverse have been the routes used by the stream in the past to reach the sea. For a long while, in historic time, the Colorado kept generally to a well-defined main channel leading nearly southward into the Gulf of California; but even so it occasionally broke through the dike, of its own building, which has separated it for centuries from the vast basin which used to be the head of the Gulf of California. This excess water has uniformly flowed northward toward the Salton Sink. These digressions were of little concern to man until steps were taken to develop the Imperial Valley for agricultural purposes—then the inundations became menaces to life and property.

**The same bridge after the Colorado had returned to normal. The river then had a maximum flow of but 20,000 second-feet.**

### *Man Settles in the Valley*

We need not ponder the questionable wisdom of those adventurous persons who early urged that the Imperial Valley could be turned into a vast farm if water could be distributed to its arid but fertile soil by tapping the Colorado River. Dr. Oliver M. Wozencraft may properly be called the pioneer in this movement, because he urged such a project after he had crossed the desert lands of the Salton Sink in 1849. Even so, it was not until the spring of 1900 that the scheme took final shape when a contract was awarded for the digging of a canal that would link the Colorado River with the dry bed of what has become known as the Alamo River—the latter taking the water delivered to it by the Imperial Canal and leading the vitalizing flood northward for distribution in the Imperial Valley. That epochal work was done by George Chaffee; and in less than two years fully 400 miles of main canals and laterals were dug under his energetic direction.

The fertile soil, the water, and the abundant sunshine together wrought wonders after seed were sown; and the amazing agricultural potentialities of the region were quickly established beyond question. The problem that soon became pressing was the security of that ambitious venture, because the participants had put themselves in a depression far below sea level and were drawing water from a river whose unstable course flowed high above the sea level. How long could the people so situated count upon holding the Colorado at bay when that capricious stream was on the rampage?

At the start, the development company led water from the river through headworks, known as Hanlon's Heading, constructed on the west side of the stream just above the international boundary; and shortly afterwards Sharp's Heading was built in Mexico just below the boundary—the idea being that water could be delivered to the Imperial Canal by either heading. Knowledge of the success of the first farmers of the land drew thousands of other settlers to the basin, and this influx led promptly to demands for larger and larger quantities of water. Difficulties had been encountered in keeping the original heading open—mainly because of the rapid accumulation of silt; and the canal operators were forced to cast about for some other means of providing needed water for the cultivated lands. Something had to be done and done quickly; and as the company was hard-pressed for funds, its management restorted to a hazardous expedient.

**Top—Relief map of the Imperial Valley, Salton Sink, and a part of the delta formed by the Colorado. The course of the river below Yuma has changed since this map was made in 1905. Bottom—Flood waters converging into a definite channel and forming falls as they cut backward upstream.**

### *Man Tempts the Colorado*

Against the advice of competent engineers a third heading was cut in the bank of the river a short distance below Sharp's Heading—this new heading affording a short and quick descent from the Colorado to the Alamo. This was done in October, 1904, and the cut was from 40 to 50 feet in width and ranged from 6 to 8 feet in depth. The excavation was made directly in the mud of the river bank; and that opening permitted a relatively small amount of water to flow into the basin for the immediate relief of the famers. What followed was thus summed up by F. H. Newell, in 1907, when Director of the United States Reclamation Service:

"The California Development Company did not have approved plans or funds available to build headworks in this opening, and it was assumed that, with ordinary care and watching, the channel could be kept open just sufficient to allow the needed amount of water to pass out from the west bank. With the next rise in the river, however, the fears of the engineers were fulfilled. Following a capricious mood, the river concluded to go down the easy channel toward the Alamo and sent from day to day an ever-increasing flood, rapidly eroding the channel. This continued until, in the spring of 1905. the entire river was passing by an abrupt turn to the westward down the Alamo channel, spreading out over the low ground and ultimately converging toward Volcano Lake or northerly into the New River and the Salton Sea. The old channel of the river, where it formed a part of the international boundary and at points below, soon became completely dry and rapidly assumed the ordinary appearance of the alluvial desert."

The flood waters of the Colorado overflowed the natural bed of the Alamo and spread out across the desert—in some sections the inundation covering the land as far as the eye could see. It was flowing over soil built up either by the sweep of winds or by the river during some of its previous incursions into the basin, and the ground was, therefore, of a character that could be cut away quickly as the flood turned into numerous separate streams which gathered headway as the increasing slope of the basin added to their velocity. These streams scoured out channels in the soft earth which, at points, were quite 1,000 feet in width, a good many feet in depth, and filled with turbulent torrents.

Following a characteristic tendency of streams flowing in yielding ground, the flood waters cut continually backward toward their source, forming deep and steep-sided gulches—the runaway waters leaving the lands that were undamaged remote from the supply of water for their growing crops. But the worst that was apprehended was that these flood streams might continue their back-cutting until they reached the Colorado, and there destroy the remaining headworks and turn the whole volume of the river down into the Salton Sea. It was even feared that the back-cutting might then travel up the Colorado and undermine the Laguna Dam, which was at that time building above Yuma!

When the trouble-making third heading was cut in October, 1904, the surface of the water in Salton Sink stood at 273.5 feet below the mean sea level. By the time the break at that heading was closed in February, 1907, the water in the sink had mounted to a point 201 feet below the sea—that is to say, it rose a total of nearly 73 feet and increased the superficial area of the so-called Salton Sea to 298,240 acres. Had the river not been balked and turned back into its established course, after months and months of desperate work, its entire flow would have continued to pour into the basin. A simple computation, based upon the annual influx, would disclose how soon the Colorado could reflood the area which it had capriciously isolated in the remote past. At the rate of evaporation normal to the region, centuries would probably pass before the Imperial Valley would again be dry.

### *Battle to Shut Out the River*

During the anxious period between the development of the dangerous break and its closure, those engaged in battling with the river believed more than once that they were engaged in a forlorn hope; but they held on grimly to the task set them. Almost at the instant that the channel was confined to any promising degree, the swift currents began burrowing and undercutting the obstruction of man's making. It is said that piles 70 feet in length were dislodged and swept away as fast as they could be driven into the water bed. Indeed, any narrowing of the channel by the placing of an obstruction merely increased the velocity of the flow and intensified the stream's erosive strength so that it cut enormous gaps in the space of only a few hours.

In the end, the fight was won by the engineering force of the Southern Pacific Railroad whose tracks were several times relocated in anticipation of further flooding of the Salton Sink and who took over the task of saving the Imperial Valley from inundation after the development company proved unequal to it. Again we shall quote from F. H. Newell's absorbing description of the work:

"Added to the unfavorable character of the bed and banks was the fact that the river seldom remained quiet for any considerable length of time. It was subject to short violent floods, especially from its tributary, the Gila. These, occurring at a time when the work was in a critical condition, quickly rendered useless the efforts of the constructors. The method finally adopted for turning the stream was one whose success depended upon having at hand a large railroad equipment and an enormous amount of material which could be quickly transported. . . . . . . . . The chief difficulty was to secure a sufficient supply of stone fast enough to fill the gaps as they were washed out. In some cases trains of flat cars loaded with stone were brought from a distance of 400 miles. . . . . . . . . As the rock heap rose gradually it checked the river, causing it also to rise higher and higher and to cascade over the pile of stone. Riffles were caused, and undercutting of the lower slope, or of the rock heap, allowed it to settle and the stones to roll downstream. All of this undercutting and settling had to be made up and overcome by the rapid dumping of other large stones. It was necessary to raise the river bodily about 11 feet. As the water rose and became ponded on the upper side of the rock heap, trainload after trainload of small stone and gravel from the nearby hills was dumped to fill the spaces between the large rocks." When hydraulicked with silt and covered with earth and gravel, the dike became effective against seepage. The breach was closed and the river forced back into its channel in February of 1907—that victory costing about $2,000,000.

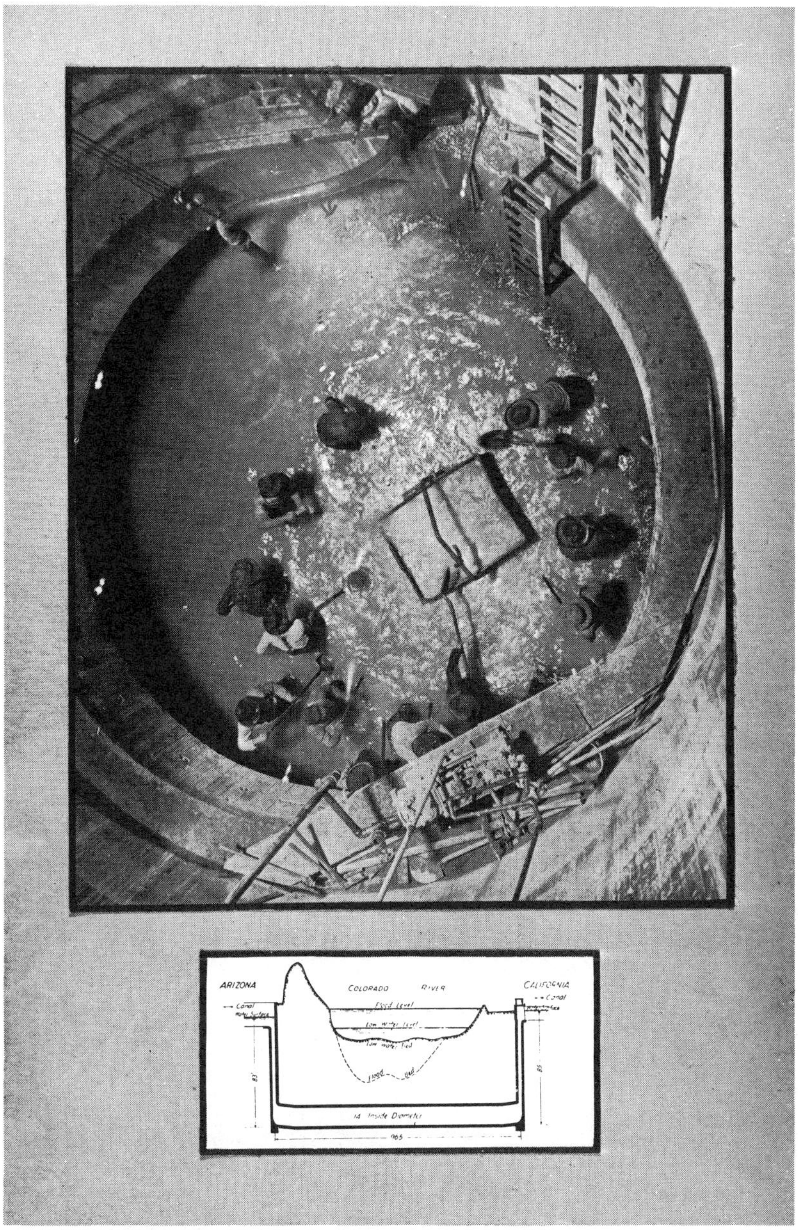


**Top—Looking down the Arizona shaft of the Colorado River siphon. Bottom—Cross section of the siphon. Note the depth to which the river scores its bed during a flood period and upbuilds it again with silt as the waters drop and resume their normal quieter flow.**

### *River Renews the Fight*

The Colorado was merely halted but not whipped. Raising its bed with deposits of silt at the rate of something like 10 feet annually, it was ready for another offensive in 1909; and the point of its attack then was only a short distance below where it had breached the defense of the Imperial basin in 1905. The river forced its way through the dike 29 miles below Yuma, and turned its flood westward through Bee River into Volcano Lake—hastily built levees sufficing to keep the stream from reaching the Alamo and the New River drainage basins. After pouring an enormous amount of silt into Volcano Lake, the Colorado carried its flood waters to the channels offered by the Hardy and the Pescadero rivers and thence onward to the Gulf of California. The rapidity with which the Colorado silted Volcano Lake made the task of rearing a levee there a desperately difficult one. However, the engineers were able to upbuild the dike fast enough to keep the crest above the rising water. Two years later the Federal Government appropriated $1,000,000 to relieve the menace at that point.

To date, approximately $7,000,000 has been expended in efforts to protect the Imperial Valley from inundating floods; and the silt carried into the irrigating channels presents a continual problem. The deposits must either be dredged out, to keep the channels at a given level, or dikes must be continually raised to prevent the water from overrunning the lands as the silt shallows the irrigating arteries. It is said that the disposal of silt in the Imperial Valley entails upon the district a yearly outlay of $1,400,000.

### *Man's Success in the Valley*

What has been told about the origin of the Imperial Valley and the uncertainties attending the Colorado when on a flood-period rampage should make reasonably clear the menace to which dwellers in the basin are continually exposed as long as there is no flood control of the river. Since the first farms were irrigated in that desert region, about 29

**Laguna Dam and sluice gates viewed from the California side of the Colorado.**

years ago, something like 500,000 acres have been brought under cultivation. Numerous towns and lesser communities have come into being; and the population of the valley at the present time probably exceeds 75,000 persons.

As might be expected, the summers are hot and the winters are mild. In the Imperial Valley the principal crops are lettuce, peas, spinach, and other vegetables; and a large acreage is given over to the growing of cantaloupes, grapes, strawberries, and citrus fruits. Because of the warmth of the climate, farmers in the Imperial Valley can produce these commodities at a very early season; and quantities are shipped to northern and eastern markets. That is to say, the foregoing foodstuffs are not generally obtainable at the same time of the year in other sections of the United States. The value of the crops produced annually in the Imperial Valley is close to $100,000,000.

Manifestly, there are substantial economic reasons for protecting this fruitful area which has been brought to its present status by settlers of great courage and abounding determination to win success in the face of grave natural odds. As matters stand now, the river may at any flood time break from its shifting and uncertain channel and turn its full volume into the Imperial Valley; and such an occurrence could not be likened to the inundating of other regions that would soon emerge from the flood waters. Should the Colorado break into the valley, and no means prove effective in an effort to turn the river back into its accustomed channel, then the influx would produce a permanent inundation because there would be no outlet for the imprisoned flood.

In lesser degrees, reclamation projects at Palo Verde, on the California side of the river and at Yuma, on the Arizona side, will be beneficially affected by any control of the flood waters of the Colorado; and it should be manifest that the well-being and security of life and property on any of these regional undertakings will hinge to a large extent upon the completion of the Hoover Dam. In short, the builders of the Hoover Dam are racing against time and the uncertain moods of America's wonder river.

**Boats were the only means of traversing Main Street in Yuma when the 1916 flood was at its height.**

U. S. Bureau of Reclamation

**Area immediately above the Hoover Dam site, seen from Monument Pass, Nevada, that will form part of the vast reservoir which is to hold more than 30,000,000 acre-feet of water.**

# America's Wonder River—the Colorado

## *Work on the Hoover Dam Now Underway After the Disposal of Many Puzzling Questions*

By R. G. SKERRET

IN curbing the Colorado by means of the Hoover Dam, the Federal Government is essaying the greatest task in hydraulic engineering that it has taken upon itself since digging the Panama Canal. In beginning this momentous undertaking, both engineers and constructors have joined battle with inexhaustible nature; and they have willed to bring under subjection a stream with a violently riotous record that reaches unbrokenly into the dim past. The work in hand is not only superlatively big in the matter of dimensions but vast in its economic potentialities and significance.

The mere start of the Hoover Dam crystallizes agitation for the control of the Colorado that dates back more than a quarter of a century. And that we may have a fair conception of the undertaking and the steps by which it has been brought to its present pass, it might be well to summarize the preliminaries that smoothed out difficulties of various sorts and that provided a firm basis for the consummation of a long-desired relief.

The Hoover Dam project as now planned calls for the ultimate expenditure of $165,000,000, including the construction of an appurtenant undertaking known as the All-American Canal. The total sum will be made up of the following four items:

| | |
|---|---|
| Dam and reservoir.......... | $70,600,000 |
| 1,000,000-hp. development .. | 38,200,000 |
| All-American Canal ........ | 38,500,000 |
| Interest during construction.. | 17,700,000 |
| Total................ | $165,000,000 |

The All-American Canal will, when completed, take water from the Colorado at a point above the Mexican boundary, and is designed to supply the needs of the Imperial Valley and the Yuma project. It will be 75 miles in length, and will be notable for a number of reasons. We shall not now touch upon this phase of the control of the Colorado River—the matter of immediate interest is the Hoover Dam and its associate features.

### *Colorado River Compact*

The building of the Hoover Dam might still be something for future decision had not the Colorado River Commission been called into existence in 1921. That commission was made up of representatives from Arizona, California, Colorado, Nevada, New Mexico, Utah, and Wyoming. Herbert Hoover, then Secretary of Commerce, was the chairman of the commission. The commission was brought into being when the several states immediately concerned had advanced numerous sincere but very conflicting claims; and it was manifest that some representative body would have to take all these claims under consideration

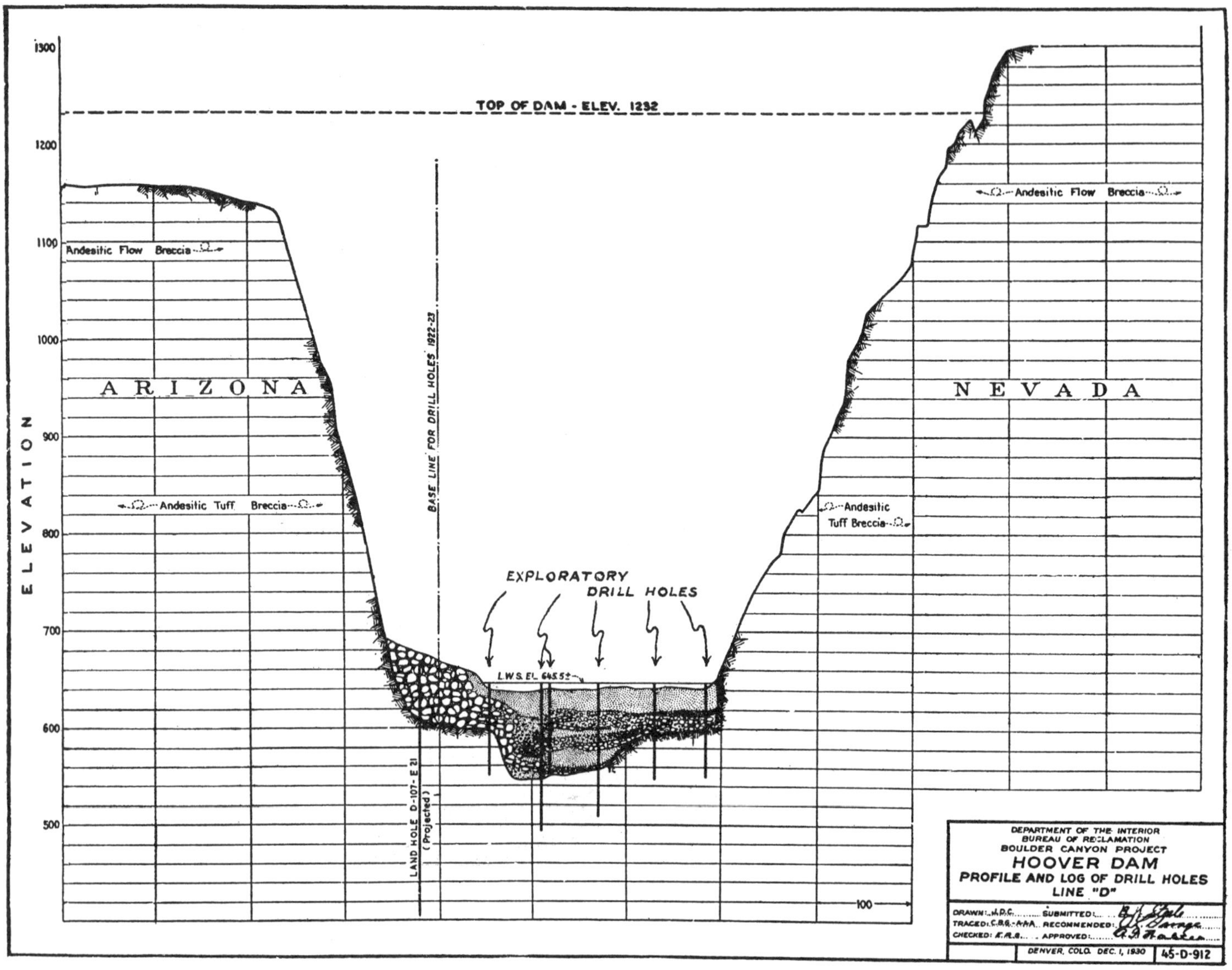

U. S. Bureau of Reclamation

**Black Canyon before work was started on the Hoover Dam project. The site is just at the far bend in the river and was chosen only after extensive investigation had revealed its outstanding advantages.**

and, if possible, bring about accord. The compact adopted at Santa Fe, N. Mex., on November 24, 1922, bears testimony to the broad-minded spirit of give and take that actuated the commissioners and enabled them to arrive at a happy understanding within a period of less than a year. Never before had more than two states thus adjusted their contending claims; and the compact is, therefore, of unusual significance. Mr. Hoover's tact, patience, diplomatic persuasiveness, and recognized engineering skill had much to do in promoting the outcome.

The commission earnestly recommended and urged the early construction of works in the Colorado River to control floods and permanently to avoid that menace. Before actual steps could be taken to that end, however, certain practical questions had to be settled. It was evident that Congress could not be asked to appropriate funds until the best site for a suitable dam were chosen and the size and the cost of such a structure predetermined. Indeed, conclusive information was lacking about some of the canyon sections of the river where it might be found desirable to erect the controlling dam. Accordingly, experts of the United States Geological Survey were detailed to make the needful examinations and to gather certain indispensable hydrographic and geologic data.

Because of physical and climatic conditions, the far-flung region traversed by the Colorado River system may properly be divided into three distinctive sections. The river basin or watershed formed by parts of Colorado, New Mexico, Utah, and Wyoming lends itself to the utilization of the stream's water for irrigation and for power development in about equal divisions. The middle area contiguous to the river largely consists of a 500-mile, high-walled canyon region–the stream dropping a total of about 3,000 feet in that distance. Throughout this area but very little water drains into the Colorado. Plainly, the canyon section is outstandingly a source of potential power and virtually unable to make more than extremely small use of the river's water for irrigation. The lower reaches of the Colorado, lying in the southern half of Arizona and the southeastern part of California, constitute an area of 69,000 square miles; and while the river throughout this region drops about 500 feet in 300 miles still it is preeminently suited to agricultural operations. Thus, it is manifest that the Colorado River Commission did wisely in dividing the whole watershed into two main divisions or basins. The Upper Basin embraces parts of Arizona, Colorado, New Mexico, Utah, and Wyoming lying above Lees Ferry, while the lower Basin is composed of those parts of Arizona, California, Nevada, New Mexico, and Utah below Lees Ferry.

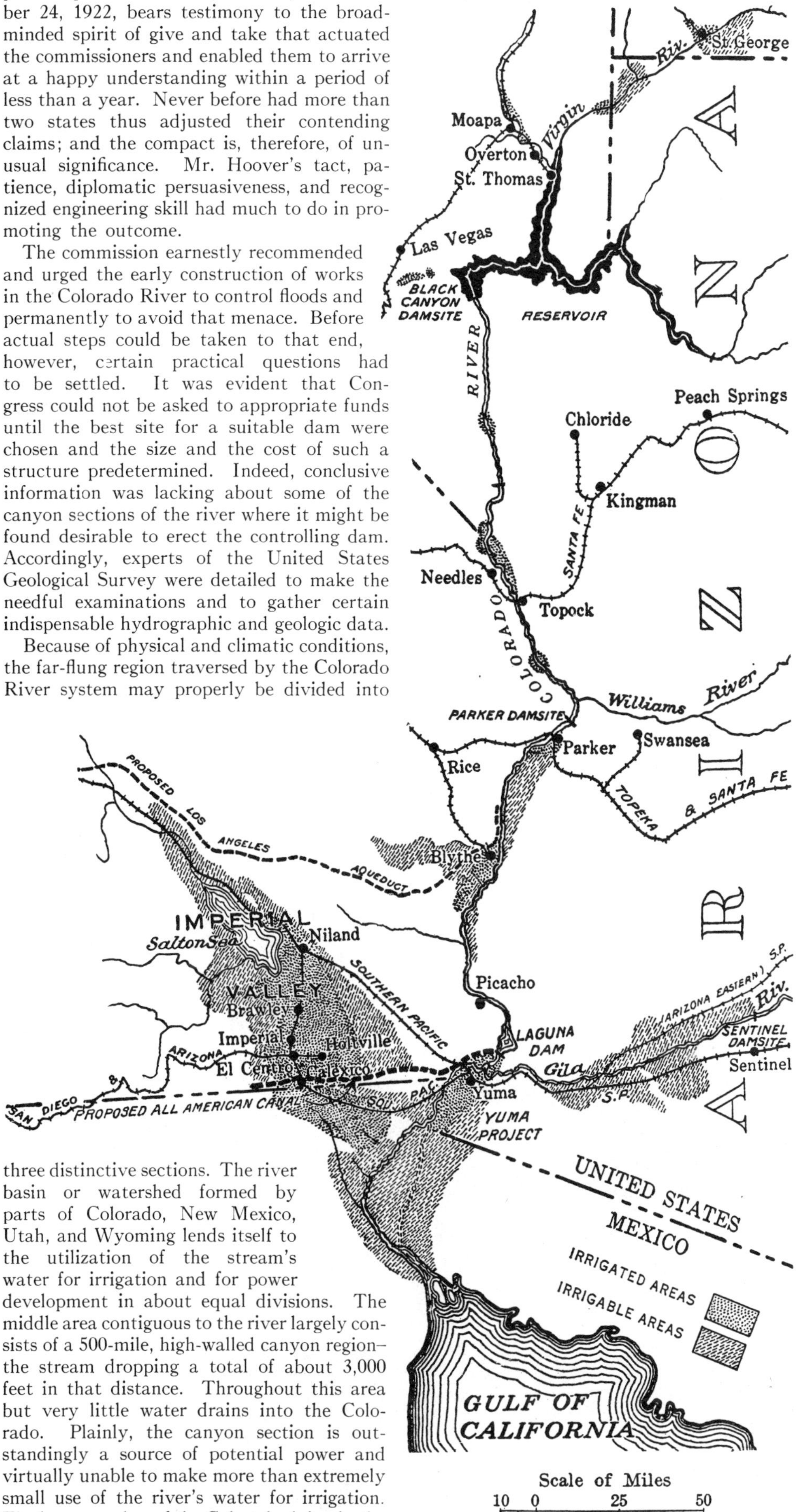


The natural flow of the Colorado River averages annually 20,000,000 acre-feet; and about one-third of this flow is now utilized. Present users in the Lower Basin take about 3,700,000 acre-feet. By the compact, this region is awarded a total yearly flow of 8,500,000 acre-feet—more than double existing needs. The Upper Basin was awarded 7,500,000 acre-feet, also more than double its present needs and ample to serve quite 3,000,000 acres of additional cultivated land. Besides these allotments, there remains an unapportioned measure of 4,000,000 acre-feet of water for future division as unforeseen demands may arise. So much for the wise and equitable apportionment of the water made by the commission.

### *Choice of Dam Site*

The steep gradient of the Colorado in the canyon section made logical the building of a control dam therein. The problem then was to find the place within that section where physical conditions would lend themselves best to the rearing of such a structure and where such a structure would be able to impound the largest volume of water. Furthermore, the dam had to be so situated that the associate power plants would be within relatively easy reach of markets for the electricity generated there. Consideration had to be given to the distances within which current could be transmitted without encountering prohibitive losses. The importance of this phase of the problem becomes apparent when it is realized that the sale of power is counted upon to repay the outlays incident to the building of the Hoover Dam. Indeed, the undertaking would probably not have been started had this method of refunding not been assured by contracts made with prospective users.

The responsibility for the choice of the dam site was placed upon the Colorado River Board on the Boulder Dam Project. The board was called into being agreeably to a joint resolution of Congress approved May 29, 1928. It was composed of eminent engineers and geologists; and its report was made towards the end of November following. Two sites were considered—one in Boulder Canyon and one in the upper part of Black Canyon; and the Black Canyon site was chosen. The reasons for the selection are thus summed up by that body:

"In general, geologic conditions at Black Canyon are superior to those at Boulder Canyon. The Black Canyon site is more accessible, the canyon is narrower, the gorge is shallower below water level, the walls are steeper, and a dam of the same height here would cost less and would have a somewhat greater reservoir capacity. The rock formation is less jointed, stands up in sheer cliffs better, exhibits fewer open fractures, is better healed where formerly broken, and is less pervious in mass than is the rock of the other site. The Black Canyon rock is not so hard to drill as that of Boulder Canyon; and it will stand better in large tunnel excavations with less danger to the workman.

"There is no doubt whatever but that the rock formations of this site are competent to carry safely the heavy load and abutment thrusts contemplated. It is well adapted to making a tight seal and for opposing water seepage and circulation under and around the ends of the dam. It insures successful tunneling, and, so far as the rock is concerned, the general safety and permanence of the proposed structures." Exploratory drilling, carried to a depth of 557 feet below the low-water surface of the rock, did not penetrate any other formation.

While the foundation rock upon which the dam will rest is a volcanic breccia or tuff, derived from probably repeated eruptions, still time has worked a transformation that has cemented the breccia into a tough, durable mass of rock capable of withstanding the weather and of resisting erosion exceptionally well. There is every indication that volcanic action ceased thousands of years ago; and the district, today, is virtually free from the effects of recurrent earth movements in that part of the country.

The board realized that the dam then contemplated would have to be much higher than any other dam built or projected, and was keenly alive to the fact that the structure would have to be correspondingly strong. Failure would let loose suddenly an enormous body of water which, sweeping unimpeded down the river channel, would probably destroy Needles, Toprock, Parker, Blythe, Yuma, and obliterate the levees of the Imperial district—creating a channel into the Salton Sea that would in all likelihood be so deep that it would be impracticable to turn the Colorado River again into its normal course. As the board expressed: "To avoid such possibilities the proposed dam should be constructed on conservative if not ultraconservative lines." That recommendation has been observed in every phase of the design of the Hoover Dam.

### *What the Dam Must Do*

The Hoover Dam may best be described as a titanic wedge of concrete that will be rigidly secured to the rock underlying the normal water bed and anchored to the towering flanking walls of the relatively narrow gorge. From base to crest the structure will have a height of 730 feet; and up and down stream the base will have a spread of 650 feet. Along the crest the dam will have a width of 45 feet and a length of about 1,180 feet; and it will be made up of substantially 3,500,000 cubic yards of concrete. The foundation will be bound to the supporting bedrock at a depth of 150 feet below the low-water level of the river. Expressed otherwise, the dam will replace rock worn away by the Colorado in its age-old rush seaward.

The function of the dam is a fourfold one: It is to serve for flood control; for the regulation of irrigation flow; for the development of power; and for the storage of silt. The basin created by the Hoover Dam will have a capacity of 30,500,000 acre-feet of water, or water enough to cover to a depth of 1 foot the combined land areas of Maine, New Hampshire, and Vermont. The lake above the dam will extend upstream for a distance of 115 miles and at its widest point have a breadth of eight miles. It will be the biggest of man-made reservoirs, and will be able to store nearly seven times as much water as the world-famous Assuan reservoir, on the Nile, after the heightening of the Assuan Dam has been completed. Expressed in another way, the Hoover Dam basin will have sufficient capacity to hold substantially a two years' flow of the Colorado!

### *Annual Flow of River*

Averages are difficult to arrive at in the case of the Colorado's flow. At times the discharge is less than 3,000 second-feet, while on one recorded occasion the outpouring mounted as high as 300,000 second-feet! Careful records made over a period of some years have given an average annual flow of 17,300,000 acre-feet. The point to be emphasized is that the Colorado in any of its flood periods presents a problem for the hydraulic engineer; and the Hoover Dam has been so designed that it can stand rigidly, with a very large factor of safety, against on-rushing torrents when the river is in a tempestuous mood. When the water is at its maximum height in the lake it will have a depth of 580 feet.

Next to controlling the flood flow of the Colorado, the reservoir created by the Hoover Dam will, because of its stupendous capacity, make it feasible to release continually into the river channel a volume of water ample enough to meet the irrigation needs of the farmlands on the Yuma project, in the Imperial Valley, and on those lesser areas in the lower reaches of this variable stream. So much stress has been laid upon flood control and the ravages wrought by the river when on its annual rampages that most people unfamiliar with the Colorado assume that it always is of sufficient volume to meet the demands of dependent agriculturists. This is not true; and there are years when there is an actual deficiency in vitally necessary water. Therefore, the conservation and control which will be made possible by the completion of the Hoover Dam will prevent these serious fluctuations and assure, from season to season, the water that may be required for the growing of crops and other essential purposes. The foregoing services are the primary ones of this magnificent project. Now let us consider briefly the two remaining objects of the undertaking—one of which is to make the venture a self-paying proposition.

U. S. Bureau of Reclamation

**Rounded cobbles embedded in ground-up lava on the Arizona wall of Black Canyon. Time has made the formation a strong and solid one.**

U. S. Geological Survey

**Left—
Radio played a helpful part in the last survey of Black Canyon.**

**Right—
Drill boats making exploratory borings at the dam site.**

**Bottom—
Making the final survey of Black Canyon.**

### *Generation of Power*

Associated with the Hoover Dam will be power plants equipped with groups of great hydro-electric generators that will have a combined rated output of 1,000,000 hp.—one-fourth of all the power that might be developed by a series of stations placed at strategic points in the canyon section of the Colorado. The turbines will be operated under an average head of 520 feet, and they will have a firm or continuous output of about 660,000 hp. The price of this firm power will be at the rate of 1.63 mills per kilowatt-hour. During periods of high water, when the operating head may be as much as 582 feet, the dynamos will generate their full power; and this seasonal current will be sold for 0.5 of a mill for each kilowatt-hour. Cheap power will be a great boon to industrial activities of all sorts, and especially to mining enterprises in California, Arizona, and Nevada lying within a radius of economical distribution.

As Dr. Elwood Mead, Commissioner, Bureau of Reclamation, has pointed out: "The power and water income from the contracts already signed will, in 50 years, bring an income of $373,500,000. Of this, the United States will receive $228,260,000 to repay money advanced, with interest. Arizona and Nevada will each receive $31,235,000; operation and maintenance will absorb $16,120,000; and there will be a surplus of $66,650,000, which will be the net profit of the Government for going into this enterprise. This sum will be disposed of as Congress may hereafter direct." This statement will answer the questioning taxpayer.

### *Storage of Silt*

As had been well and wittingly said, the Colorado when in flood is a stream too thick to drink and too thin to plow. This description is based on the fact that the river will pour into the Hoover Dam reservoir approximately 100,000 acre-feet of mud every twelvemonth; and the major part of this alluvium will find a resting place on the bottom of the vast basin. As the reservoir is not designed to purge itself, the silt will accumulate year after year; and, because of this, a definite volume of the bottom part of the basin will serve for silt storage. Allowing for an annual deposition of as much as 137,000 acre-feet, still at the end of the first 50-year period there will be available for water storage something like three-fourths of the total capacity of the reservoir. It has been computed by experts that it would require a span of 222 years to completely fill the basin with silt—assuming that no other dams had been built in the interval above the Hoover Dam to catch some of the silt.

The arresting of silt in the Hoover Dam reservoir will, in the course of something like ten years, work a marked transformation in the condition of the water flowing from the dam downstream. The assumption is that within that span the river below the dam will be cleared out and that the water will flow over a clean sand or gravel bed. This will be of great benefit to the areas now drawing off water for irrigation, and also to Los Angeles and the associate group of cities which will tap the Colorado for vitally necessary water.

### *Work Started on Project*

On April 20, 1931, a contract for the construction of the Hoover Dam was signed by the Secretary of the Interior, Ray Lyman Wilbur, and by W.A.Bechtel, now president

Top-How Hoover Dam and its appurtenant power plant will look when ready for service. Bottom—Model of Hoover Dam and its associate features. The lake created by the dam will be 115 miles long and capable of impounding a maximum of 30,500,000 acre-feet of water.

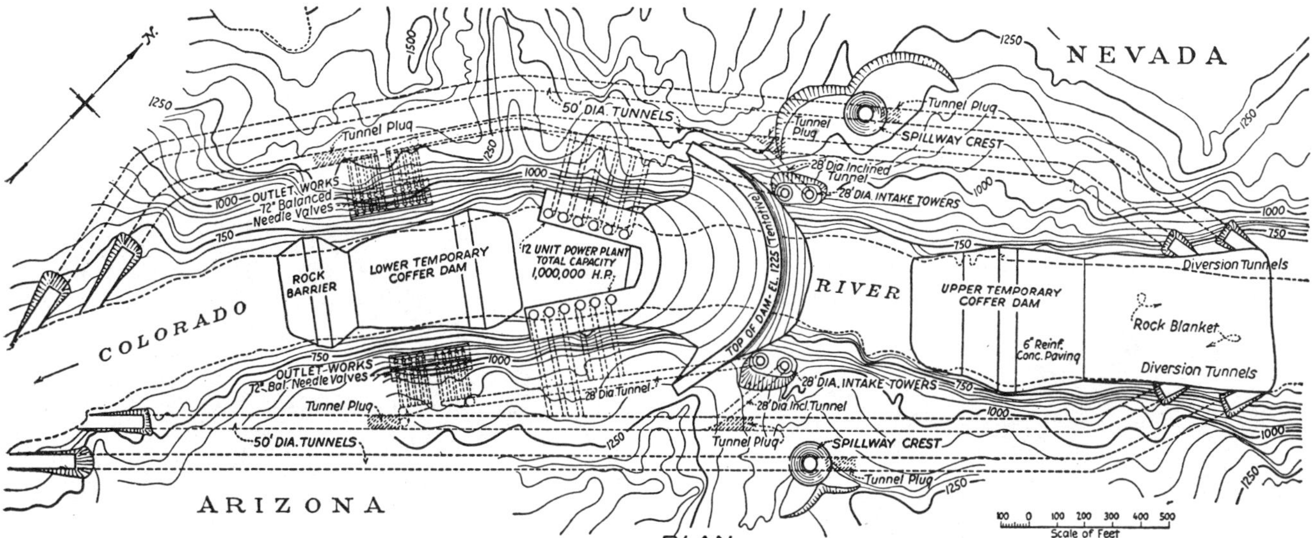


PLAN

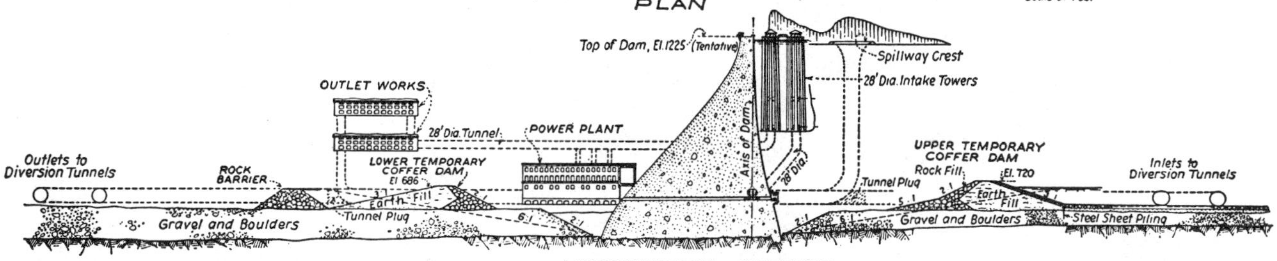


LONGITUDINAL SECTION

of the Six Companies Incorporated. Six Companies Incorporated contracted to build the Hoover Dam, the power plant, and certain associate works for the sum of $48,890,995. That act brought to a close long years of waiting and, to that extent, relieved the anxieties of our people in the lower sections of the river where there has been an ever-recurrent fear that the Colorado would break loose and overrun and ravage the contiguous cultivated lands

With the project actually underway, it is now possible to look backward and to realize that much has probably been gained by delay. Both the engineers responsible for the present magnificent scheme and the contractors engaged in carrying it out are richer in their knowledge and their experience than they would have been had they tackled a task of like proportions twenty years ago. Also, makers of machinery have, in the intervening decades, devised equipment of different kinds especially fitted to deal effectually and rapidly with the heavy and complex work involved. The contractors are called upon to rear a mountainous obstruction athwart the erratic flow of the Colorado and, at the same time, to divert that stream temporarily around this rising bulwark no matter what may be the state of the stream the while!

### *What the Contractor Must Do*

The Hoover Dam site is about 25 miles southeast of Las Vegas, Nev., and situated in what was an utterly isolated desert region before steps were taken to provide means of transportation from the nearest railroad and to create accommodations for the army of workers that will be engaged on the job for quite 6½ years—the period during which Six Companies Incorporated pledges itself to complete its part of the great undertaking. Interesting as it would be to tell how these preliminaries were carried out and how a sizable town—Boulder City—was called into being in the midst of that hot and arid region, space compels us to confine ourselves strictly to the work involved in connection with the dam and its immediately appurtenant features.

Before any part of the dam can be started, the contractor must drive four tunnels which will be 56 feet in diameter and average more than 4,000 feet in length. Two of these are being driven on the Nevada side of the Colorado and two on the Arizona side—the tunnels piercing the rock at an elevation close to that of the low-water level of the river. These tunnels are counted upon to carry the entire volume of the flow at any likely stage. Upstream above the dam site, and just below the intake ends of the tunnels, a great cofferdam, 80 feet in height, will be constructed. This will block the normal course of the river and divert its flow into the diversion tunnels. Just upstream from the outlet ends of the tunnels, and below the dam site, will be built another massive cofferdam that will prevent the Colorado from encroaching upon the space between the two cofferdams. This space will be unwatered so that the river bed can be bared preparatory to removing the accumulated sand and gravel down to bedrock, from 80 to 100 feet below the low-water level. The excavation for the foundation of the dam will necessitate the removal of substantially 7,000,000 cubic yards of material.

In driving the diversion tunnels, which, with the cofferdams, will be completed in about two years, the contractor will drill, blast, and muck nearly 2,000,000 cubic yards or rock. Something like three-fourths of this rock will be used to stablilize the two cofferdams. With the cofferdams built and with the tunnels driven and lined with 3 feet of concrete, the contractor can start work on the dam itself, and he will then bring into play a multiplicity of facilities that he has been making ready and assembling in the meanwhile. As time goes on, it is our purpose to keep our readers posted on the progress of the work, and, therefore, we shall not confuse by entering into details now.

The building of the Hoover Dam involves a battle with climatic conditions that are extremely severe and trying during the warm months of each year. Again, we quote Doctor Mead: "The summer wind which sweeps over the gorge from the desert feels like a blast from a furnace. At the rim of the gorge, where much of the work must be done, there is neigher soil, grass, nor trees. The sun beats down on a broken surface of lava rocks. At midday, they cannot be touched with the naked hand. It is bad enough as a place for men at work. It is no place for a boarding house or a sleeping porch. Comfortable living conditions had to be found elsewhere, and these are found on the summit of the Divide, five miles from the dam. Here there is fertile soil; here winds have an unimpeded sweep from every direction; and here there is also an inspiring view of deserts and lonely gorges and lofty mountain peaks."

When the dam is an accomplished fact, when the pent-up waters are available for controlled irrigation and for the generation of power, then that amazing structure will constitute a milestone for a civilization under which unnumbered generations will flourish. A vast new empire of industry and enterprise will center upon that towering monument to engineering skill and to the never-failing resourcefulness of the men responsible for its consummation.

**Looking upstream in Black Canyon toward the site of the Hoover Dam. The lower portals of three of the four diversion tunnels are shown. In the lower right-hand corner is the settling basin where river water is clarified before being pumped six miles to Boulder City. The dam will be built at about the point where the river disappears behind the cliff at the right. From a photograph made about December 1, 1931.**

(C) W. A. Davis, Las Vegas, Nev.

**Left—Bureau of Reclamation officials on an inspection tour in Black Canyon. Right—The first blast which was fired just above Black Canyon on September 16, 1930. Insert—Secretary of the Interior Wilbur driving a spike of Nevada silver to mark the beginning of construction of the railroad to Boulder City.**

# Construction of the Hoover Dam

*Some General Facts Regarding the Undertaking and the Men Who are Directing It*

By C. H. VIVIAN

BIDS for the construction of the Hoover Dam, power plant, and appurtenant works were opened March 4, 1931, at the Denver office of the Bureau of Reclamation. Only three bids were submitted, although contractors from all sections of the country were present and many of them had expected to bid until the time set for the opening drew near. The lowest bid, submitted by Six Companies Incorporated, of San Francisco, and subsequently accepted, was $48,890,995.50. The second bid of $53,893,878.70 was submitted by the Arundel Corporation, of Baltimore, associated with Lynn H. Atkinson of Los Angeles. Woods Brothers Construction Company, of Lincoln, Neb., and A. Guthrie & Company, of Portland, Ore., were the third bidders with $58,653,107.50.

The voluminous plans and specifications prepared by the Government engineers, and appraised as models of their kind, divided the complete work into 119 separate items for bidding. On most of these items the three bids were fairly uniform. The only outstanding difference between the two lowest bids was on the item of placing 3,400,000 cubic yards of concrete in the dam, the respective figures being $2.70 and $4.15 per cubic yard. This variation accounted for $4,930,000 of the $5,002,883.20 difference between those bids. Both concerns bid $8.50 per cubic yard for the tunnel excavation. The three largest items in the successful bid were $13,285,000 for the 1,563,000 cubic yards of tunnel excavating; $9,180,000 for placing the concrete in the dam; and $3,432,000 for lining the diversion tunnels with 312,000 cubic yards of concrete. No other single item was as much as $1,000,000. From the foregoing facts it can be seen that the construction of the dam and its associated works resolves itself into the performance of a few major tasks and a multitude of smaller but closely related ones. The successful bid was only $24,000 more than the estimated cost as computed by the Government engineers. This was considered remarkable close figuring in view of the great size of the undertaking.

U. S. Bureau of Reclamatiot

**Dr. Elwood Mead, Commissioner, Bureau of Reclamation.**

Upon the occasion of signing the authorization of the contract on March 11, Secretary Ray Lyman Wilbur of the Department of the Interior commented that it was the largest contract ever let by the Federal Government. He predicted that the consummation of the project would transform the desert region below the dam site into one capable of supporting 5,000,000 persons. "It is a satisfaction to see this great contract get under way", he said. "The Colorado River, instead of being a menace, will now be a great benefit".

Just how this project compares in size with previous ones of the same general character, may be judged by the fact that the Bureau of Reclamation, during the 29 years of its existence prior to 1931, let contracts calling for an aggregate use of 4,400,000 cubic yards of concrete, whereas 4,500,000 cubic yards will be required for this one job.

The terms of the contract, which departed

**Some leading figures in the Hoover Dam project. From left to right—Ray Lyman Wilbur, Secretary of the Interior; Walker R. Young, resident construction engineer for the Bureau of Reclamation; F. T. Crowe, general superintendent for Six Companies Incorporated; Raymond F. Walter, chief engineer, Bureau of Reclamation.**

somewhat from established practices, were considered very fair by bidders. They provide that the Government shall purchase practically all materials required, such as cement, steel, etc. This relieves the contractors from possible losses arising from fluctuations in the costs of materials. Accordingly, bidders were able to figure more closely the actual costs of doing the work. Another provision favorable to the contractors is that the Government will assume all damage which may result from floods experienced after the cofferdams and diversion tunnels are completed and accepted. This means that for more than two-thirds of the period of construction, the contractors will be relieved of loss from high water.

The entire work to be carried on under the Hoover Dam contract consists of five major divisions:

1. River diversion works, consisting of upstream and downstream cofferdams and four tunnels through rock, each 56 feet in diameter before lining with concrete and approximately 4,000 feet long.
2. A concrete arch gravity type dam 1,180 feet long on the crest and 727 feet high, with a radius of curvature of 500 feet.
3. Two spillways, one on each side of the river, each consisting of a 50x50-foot stoney gate, a concrete ogee overflow crest 700 feet long, and a concrete-lined open channel.
4. Twin outlet works of similar design and capacity, each consisting of two separate systems regulated by cylinder gates in the bottom of intake towers.
5. A U-shaped power house of concrete and structural steel immediately below the dam.

The Government will furnish all materials and equipment that become a permanent part of the completed structure, delivering it to the contractors at Boulder City, the construction town. The hydraulic and electrical machinery, equipment, and wiring for the power house will be installed by the Government, and the contractors will place concrete around such machinery as it is installed. The Government will also furnish and Six Companies Incorporated will install an inclined freight-car elevator alongside the canyon wall immediately below the power-house site on the Nevada side of the river. This will connect at its upper end with a highway from Boulder City, and will be available for transporting machinery and supplies for the power house to the canyon bottom, a vertical distance of apprriximately 600 feet. Sand, gravel, and stone for concrete will be taken by the contractors from a Government owned property known as the Arizona gravel pits located eight miles up the river from the dam site.

From an engineering standpoint, one of the most interesting clauses in the contract is that providing for induced cooling of the concrete in the dam as it is poured. The chemical reaction which takes place during the setting of concrete is accompanied by the generation of considerable heat. If large blocks are poured at a time, subsequent cooling may result in the development of contraction cracks. To promote cooling, it is customary to build large concrete dams in a series of columns of rectangular section and to leave spaces between these columns to provide for the greatest possible exposure of the faces to the air. Columns are built up in stages, the amount of concrete poured at a time being limited sufficiently to permit ample cooling before more material is placed on top of it. The spaces between the original columns are later filled with other columns, which are likewise built up progressively. Subsequent grouting helps to consolidate them into a solid or monolithic structure.

**Installing a 7,000-cubic-foot-per-minute compressor plant at lower end of the work on the Nevada side of the river.**

Because of the great mass of

**Three officials of Six Companies Incorporated who are closely identified with the momentous work in hand. From left to right—Henry J. Kaiser, chairman of the executive committee; Charles A. Shea, director in charge of construction activities; and Director H. J. Lawler.**

concrete—3,400,000 cubic yards—that will go into the Hoover Dam, however, the time required for its cooling by natural means would be unduly long. To expedite matters, it was decided to follow the columnar method of construction and to hasten cooling by circulating cold water through the concrete by means of built-in pipes. For this purpose, there will be 800,000 linear feet of 2-inch pipe or boiler tubing embedded in the concrete. This piping will remain a permanent part of the structure, and will be filled with grout under pressure.

Cold water for circulating through the concrete will be supplied by a refrigerating plant maintained by the contractors. With a thought for this latter phase of the work, Six Companies Incorporated bought air compressors for the initial or drilling phases of the operations that can readily be transformed into ammonia compressors on the ground. The compressors are Ingersoll-Rand 2-stage units driven by synchronous motors. Some are of the large Class PRE design: others are the intermediate size designated as Type XRE.

Most of the work during the first three years will consist of excavating and tunneling and making ready for subsequent operations. Placing of concrete in the dam proper is not scheduled to begin until December 1, 1934. The time allowance for completion of the contract is 2,565 days, or approximately seven years. The contractors are subject to a penalty of $3,000 for every day of overtime required to complete each of the five divisions of the work.

Six Companies Incorporated is a combination of six prominent contracting concerns operating in the West. It was incorporated in February, 1931, for the express purpose of pooling resources, experience, and personnel in so far as the Hoover Dam is concerned. For all other purposes, the several member firms have retained their individual identities and are carrying on their separate activities as before.

The firms making up Six Companies Incorporated, with the percentages of their participation in the profits or losses that may accrue from this contract, are: W. A. Bechtel, San Francisco, and Henry J. Kaiser, Oakland, Calif., 30 per cent; Utah Construction Company, Ogden, Utah, 20 per cent; MacDonald & Kahn Company, Los Angeles, 20 per cent; Morrison-Knudsen Company, Boise, Idaho, 10 per cent; J. F. Shea Company, Portland, Ore., 10 per cent; and Pacific Bridge Company, Portland, Ore., 10 per cent. These companies had, collectively, completed $409,000,000 in contracts up to March, 1931, and had underway other work totaling more than $30,000,000. All told, they had carried out 3,024 individual contracts, of which 178 were for Government work, and had 69 additional contracts in hand at the time they were awarded the Hoover Dam undertaking.

Of these member companies, the Utah Construction Company is perhaps the best known, as it has fulfilled contracts in virtually every state in the western third of the country. Since its formation, in 1900, it has done approximately $200,000,000 worth of contract work—chiefly railroad construction, but including also considerable irrigation and reclamation work. The Pacific Bridge Company was organized in 1869 and has specialized in bridge building, particularly in underwater foundation work. W. A. Bechtel Company has been operating for seventeen years, during which period it has completed approximately $30,000,000 worth of railroad, dam, and general construction contracts. Kaiser Paving Company, Ltd., dates its corporate existence back to 1913, and has specialized in paving contracts. MacDonald & Kahn Company, made

**River camp of Six Companies Incorporated, at Cape Horn, above Black Canyon, where some 400 workmen make their homes.**

**Left—A section of "Ragtown", a riverside mushroom settlement above Black Canyon. Right—Black Canyon, viewed from above. Bottom—A pack train of one of the Government's early surveying expeditions along the Colorado.**

up of Alan MacDonald and Felix Kahn, has a record of $75,000,000 in contracts fulfilled, consisting chiefly of building construction on the Pacific Coast. The firm was formed in 1920. The Morrison-Knudsen Company, composed of H. W. Morrison and M. H. Knudsen, has been in business since 1912 and has completed $30,000,000 worth of contracts covering the building of roads, railroads, dams, and other construction work. J. F. Shea Company has accounted for some $40,000,000 in constracts since its organization in 1914. It has done a general contracting business, with special attention to tunnel work.

It can be seen from the foregoing brief resume of their past activities that the component members of Six Companies Incorporated are admirably equipped to supervise and perform the many and diverse tasks that enter into the building of the Hoover Dam. Their combined experience covers every phase of the work now in hand. Some of the things to be done in Black Canyon are on a larger scale than ever before attempted, but among the personnel of the contractors are men who have had ample theoretical and practical acquaintance with the problems involved.

The officers of Six Companies Incorporated are: W. A. Bechtel of Bechtel & Kaiser, president; E. O. Wattis of the Utah Construction Company, first vice-president; H. W. Morrison of the Morrison-Knudsen Company, second vice-president; Felix Kahn of MacDonald & Kahn Company, treasurer; Charles A. Shea of J. F. Shea Company, secretary; K. K. Bechtel of Bechtel & Kaiser, assistant secretary and treasurer. W. H. Wattis, elected president of the company at the time of its organization, died in September, 1931. He was for many years an officer of the Utah Construction Company, and was one of the leading construction men of the West. Directors of the company are: W. A. Bechtel, S. D. Bechtel, Philip Hart, Henry J. Kaiser, Felix Kahn, Alan MacDonald, H. W. Morrison, Charles A. Shea, E. O. Wattis, H. J. Lawler, and Guy LeR. Stevick.

While all the directors of Six Companies Incorporated visit the operations at regular intervals, the active management is in the hands of an executive committee composed of four members—Henry J. Kaiser, chairman; Charles A. Shea, director of construction; Felix Kahn, in charge of all activities of the subsidiary Boulder City Company which is charged with the feeding, housing, and transporting of the men; and S. D. Bechtel, in charge of purchasing, auditing, and warehouse activities. Mr. Shea spends most of his time on the job, and is as active as Francis T. Crowe, the general superintendent of Six Companies Incorporated, in supervising the manifold construction details involved. He acts as contact man between the board of directors and the operations personnel. In addition to the Boulder City Company, another subsidiary, the Hoover Dam Transportation Company, was formed.

The man upon whose shoulders rests the chief responsibility for carrying through this record-breaking contract is the general superintendent. Mr. Crowe is thoroughly schooled in the work he is directing, both from the standpoint of the contractors and of the Government bureau for which the work is being done. He is 49 years old, and was graduated from the University of Maine. He entered the service of the Bureau of Reclamation in 1904 as engineering aid, and was advanced until he was general superintendent of construction attached to the Denver office when he resigned in 1925. While with the Bureau he was in charge of construction of the Jackson Lake Dam in Wyoming and of the Tieton Dam on the Yakima Project in Washington, the latter an earth-and-rock fill structure 222 feet high. He served for a time as assistant to the construction engineer on the 349-foot Arrowrock Dam in Idaho. He also was project manager of the Flathead (Indian) Project in Montana. In 1925 Mr. Crowe went with the Utah Construction Company and the Morrison-Knudsen Company, jointly, as superintendent of construction. In that capacity he had charge of building the Deadwood Dam, a Bureau of Reclamation enterprise.

Mr. Crowe will "live on the job" both literally and figuratively until the Hoover Dam is completed. He established his residence at Boulder City soon after the bids were opened, and has since been there almost continually. It is not unusual to find him down in the canyon at 12 o'clock at night and again at 5 o'clock the following morning. He has surrounded himself with a group of efficient assistants for directing the tasks in hand. Most of his immediate staff were taken from the various member firms of Six Companies Incorporated. Virtually all of them have had years of experience in construction work similar to that now facing them. To Mr. Crowe and his assistants the building of the Hoover Dam is just another job—a job that differs from previous accomplishments only in point of size and time requirement.

Something has been said in previous articles of this series about the extensive investigations, by Government engineers, which led to the selection of the Black Canyon site. Any account of the building of the Hoover Dam would be unfair and incomplete if it did not pay tribute, at the outset, to the Bureau of Reclamation for its part in paving the way for the contractors. Once the dam site was chosen, an almost incomprehensible amount of detail work had to be done in the field in order that adequate plans and specifications might be prepared. As these data were obtained, their translation into printed forms necessitated an enormous amount of office engineering work.

Unfortunately, there is not available a written record of the activities of the field force. It goes without saying that many hardships were endured. Whereas the men who build the dam will live in a modern, sanitary, city, the engineers who pioneered this great undertaking had only a cluster of tents that they could call home. Deprived of all but the most primitive necessities, they traversed a country without roads and almost without shade, risked their lives on precipitous rock walls, and carried on in blistering summer heat. Dr. Elwood Mead, Commissioner of Reclamation, gives a hint of what confronted them in the following words:

"The survey of the dam site and reservoir was of unprecedented magnitude and difficulty. It involved coping with a river which, in the highest floods, rushed through the canyon with the speed of a railway train; of taking topography in more than 100 miles of canyon where precipitous cliffs 1,000 feet high and of indescribable ruggedness had to be scaled. Three lives were lost in this hazardous undertaking. Every phase of the work involved great danger, but the dimensions of the possible dam and reservoir had to be known. Then there had to be a topographic map.

"Anyone who views the canyon either from the top of the rim or from the river at the bottom has a sense of the peril and hardship involved in fixing locations and making measurements on its cliffs. To have done this work by the old methods would have delayed beginning construction six months to a year. Resort was had to aerial surveys. This involved great hardship because of the intense summer heat and in making observations at great differences in elevation".

While the Bureau has discharged with credit its great task of preparing the engineering preliminaries for this mammoth undertaking, it is still faced with the responsibility of supervising and checking every phase of the construction activities to see that the plans and specifications are faithfully followed. Space permits mention of only a few names of individuals who have been and will continue to be prominent in the Government's part in the building of the Hoover Dam.

As the Bureau of Reclamation is a division of the Department of the Interior, Secretary Wilbur is the nominal head of its activities. Having served as president of Stanford University and lived in the West for many years, he carried with him to his present office a thorough understanding of the Colorado River problem and a realization of the benefits that would result from its solution. It was through him and his assistants that the agreements and contracts were carried through which made possible the start of actual construction operations.

Doctor Mead, directing head of the Bureau of Reclamation, is declared by many to be the world's leading authority on land reclamation. A long list of completed projects in the western part of the United States bears testimony to his efforts; and he has served as consultant and director on similar work in Australia and Palestine. Like Secretary Wilbur, he is a western man, having started his engineering career as state engineer of Wyoming.

As chief engineer of the Bureau and head of its Denver office, Raymond F. Walter had active charge of the engineering staff which planned and designed the Hoover Dam and its appurtenant works. He is a specialist on problems connected with irrigation and with power development. John L. Savage served as chief designing engineer for the dam.

Walker R. Young is resident construction engineer. He designed the Arrowrock Dam, until recently the highest in the world. He was in charge of the Kittitas division of the Yakima Project, which involved a unique and very difficult task of canal construction, and directed the field work in connection with the investigations of dam sites on the Colorado River. The Bureau of Reclamation has maintained an office at Las Vegas, Nev., 30 miles from the dam site, since construction began, and will this month move into the Government administration building in Boulder City. John C. Page is office engineer there.

# THE SECOND SUMMER

DURING the first year of operations, the workers were obliged to endure many hardships, particularly because they had little protection against the blistering heat that hovers over the desert for five months each year. By the second summer, however, thanks to the efforts of Uncle Sam and of the contractors, the dam builders and their families were fairly comfortable in Boulder City, a $2,000,000 model municipality in the midst of a sandy waste. Nearly 5,000 persons were living in this de luxe construction camp in July, 1932, when the picture shown above was taken.

**This attractive Spanish-type building is the home of the directors of Six Companies Incorporated during their frequent visits to the work. It also serves as a guest house.**

# Construction of the Hoover Dam

## *How the Contractors Handled the Huge and Costly Program of Preliminary Work*

C. H. VIVIAN

DESPITE the unprecedented size of the Hoover Dam contract, the multiplicity of operations involved, and the comparative isolation of the site, Six Companies Incorporated organized its forces and entered upon its huge task in surprisingly short time. Formal notification by the Government to begin work was not given until April 20, 1931, following the actual signing of the contract by Secretary Wilbur of the Department of the Interior. The contractors did not stand on ceremony, however, and little more than a week after the opening of bids at Denver, on March 4, Supt. F. T. Crowe was on the ground recruiting a labor crew to carry on the initial stages of the 7-year job and at the same time making ready for subsequent operations. First of all he opened an office at Las Vegas, Nev., roughly 30 miles from Black Canyon. Then he visited the site where the construction town of Boulder City was to rise and started a battalion of men building quarters for the army of workers that was to follow.

The speed with which Six Companies Incorporated inaugurated work is characteristic of the zeal and industry which have since been shown. At the very outset, the gigantic undertaking was resolved into its various major factors and every phase of the work that could be started was got underway with the least possible delay. By June, 700 men were at work, and the payroll was $100,000 monthly. Almost over night the desert quiet was transformed into teeming activity. A 3-shift day and a 7-day week were put into effect as soon as the working forces had been fairly organized, and that schedule was adhered to month after month almost without a break. Neither the sizzling summer heat nor the advent of legal holidays was allowed to slow things up. As a result of this vigorous program, remarkable progress has been made during the first year of operations. If maintained, and there seems to be every reason that it will be, it is bound to bring about completion of the contract well ahead of the time limit set.

**W. A. Bechtel, president of Six Companies Incorporated.**

It is safe to say that never before has there been a project which required so much preliminary work before the real purpose of accomplishment could be attacked. While no official figures are available on the point, a member of the Six Companies organization has estimated that approximately $2,000,000 was spent before a shovelful of "pay dirt" was turned over. This huge sum, which had to be expended merely to prepare the way for performing the principal work involved, is in itself sufficient to carry out a complete contract of no mean size.

The advance guard of dam builders was greeted by a desolate region of no buildings and few roads, populated principally by little vari-colored lizards and an occasional jackrabbit. The arid climate and the sandy soil combine to limit the vegetation to little other than mesquite and cacti, and one might travel mile upon mile without seeing a tree. Roughly 1,400 feet below the general level of this barren expanse, the temperamental Colorado River, red and roily, has cut a serpentine

Three views of typical structures among the 475 buildings erected by Six Companies Incorporated in Boulder City. Top—The $50,000 hospital. Center—Two of the 172-man dormitories, of which there are eight. Bottom—A row of 3-room houses for married workers and their families.

**When Boulder City was young. Carpenters living in the tents shown above built 40 eight-man bunk houses on skids. At the left is a close-up of one of the bunk houses.**

course. At Black Canyon it is hemmed in by sheer walls of igneous rock 600 feet high: above the canyon the basin widens to a 12-mile span from rim to rim.

From the outset, there was no attempt to compromise with nature. "We must have a place to eat and sleep before we can put men out there," Superintendent Crowe stated. Accordingly, the first plans unrolled were those for a construction camp. The canyon bottom near the dame site was virtually inaccessible at the time except by boats; moreover, it was a veritable inferno in summer. The Government had wisely selected a site seven miles from the river on the Nevada side for a base of operations. Here it was that Six Companies Incorporated pitched camp and formed the nucleus of Boulder City—the fastest growing town America has known since the era of the gold rushes. During the intervening period of less than twelve months, the contractors have spent upwards of $800,000 for the construction of more than 475 buildings, practically every one of which will be torn down as soon as the dam is completed. This because Uncle Sam hopes to mold a model municipality there. Details of its physical and political make-up and the part that the Government is playing in its development will be presented in a later article. It is our present purpose to sketch, in a broad way, the steps that the contractors had to take by way of approaching the main task in hand, and to set down some of the things that were done to bring a measure of comfort to the workmen in a section of trying climatic restrictions.

At the time the Six Companies forces appeared on the ground, the Union Pacific Railroad was completing a 22-mile branch from a point near Las Vegas on its Salt Lake City-Los Angeles line to the site of Boulder City. Meanwhile, construction was in progress on both a railroad and a highway extending from Boulder City to the edge of the Nevada canyon wall above the top of the projected dam, the Government having let contracts for these in January.

The building of the 10½-mile railroad was in the hands of the Lewis Construction Company, of Los Angeles, whose bid of $455,509.50 was the lowest of the sixteen submitted. The line traverses a rough section which called for a ruling grade of 3 per cent and a maximum of 5 per cent. Its construction involved the moving of 900,000 cubic yards of earth and 202,000 cubic yards of rock, as well as the boring of five tunnel sections aggregating 1,705 feet in length and requiring the excavating of 26,000 cubic yards of rock. The tunnel work was done by Joe Gordon of Denver. The contract specified completion within 200 days from the beginning of work.

The contract for the gravel-base, oil-surfaced highway, 22 feet wide and 43,972 feet long, was let to the General Construction Company of Seattle, Wash., which gave a subcontract to R. G. LeTourneau, Incorporated, of Stockton, Calif. The work entailed the handling of 107,000 cubic yards of common excavation and 228,000 cubic yards of rock. Incidentally, something of a record was made in getting started on this work.

Two days after the contract was let it had been transferred to LeTourneau, Incorporated. Operations were started four days later, on January 28. On January 30, the Anderson Brothers Supply Company of Los Angeles had facilities ready to care for 100 men. By February 7, fifteen carloads of machinery and equipment had been delivered, including fifteen Caterpillar tractors and two Ingersoll-Rand portable air compressors.

Through the instrumentality of the Government, work was also underway on another very important medium of service to the contractors. This was a transmission line to deliver power to the site from generating stations at Victorville and San Bernardino, Calif., the latter more than 200 miles away. The contract for furnishing power was awarded jointly to the Southern Sierras Power Company and the Nevada-California Power Com pany. The line, together with a substation near the rim of the canyon on the Nevada side, was designed and constructed by the first-named company at a cost of approximately $1,500,000. Here, again, unusual speed was shown in the face of many obstacles. Despite the fact that a considerable portion of the route was across mountainous country, the 193-mile line from Victorville to the substation was put in place at the rate of 1.45 miles a day, which is said to constitute a

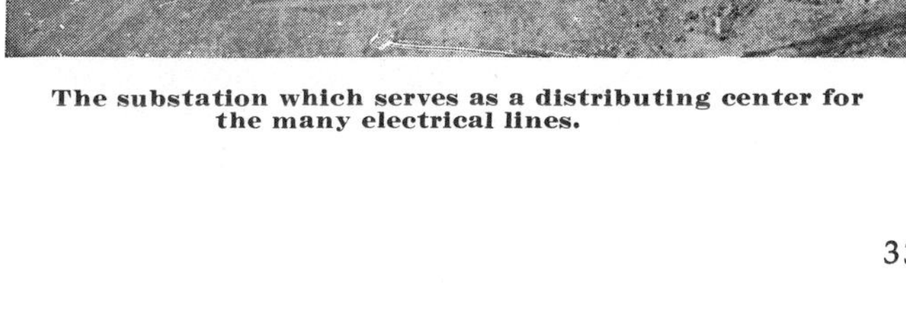

**The substation which serves as a distributing center for the many electrical lines.**

**Railroad and highway work were important and costly parts of the preliminary work. The view above shows the first tunnel holed through on the Government's railroad to the rim of Black Canyon. At the left is a scene during highway construction.**

record for such work. Field camps for from 50 to 80 men each were established at suitable intervals, and as many as five of them were maintained at a time. In some cases the trucks that delivered materials and supplies had to make their own roads; and on one occasion it was necessary to let a truck down a steep grade by means of a winch and cable. Construction activities extended over a distance of 125 miles at one time. The line consists of 2-legged, fabricated-steel towers with 34-foot, steel, angle cross arms, spaced seven to the mile. Approximately 5,000,000 pounds of steel and 1,080,000 pounds of aluminum-strand, steel-reinforced cable were used. A telephone line parallels the power line.

Construction of the substation on a high, rocky point having a steep approach was accompanied by difficulties. A compressor to furnish air for excavating the 2,100 cubic yards of rock required to be moved for the placing of foundations was packed up the hillside in sections by burros. Later a temporary switchback road, having grades up to 17 per cent, was built to permit the moving in of construction materials and station equipment. Power was turned on on June 25, beating by several days the time limit of 240 days allowed for designing and building the system. R. H. Halpenny was in charge of design and E. J. Waugh was construction engineer. Field forces on line construction were in charge of C. H. Rhudy; and H. O. Watts supervised the building of the substation. The line is insulated for 132,000 volts, but power is being transmitted at 80,000 volts. A 6.83-mile, 33,000-volt, wood-pole line was built from the substation to Boulder City, and a .73-mile, 2,300-volt line was constructed into the canyon to furnish power to the No. 1 pumping station of the water-supply system for Boulder City.

As previously written, the first concern of Six Companies Incorporated was to provide adequate living quarters for their personnel. In this connection it should be emphasized that the ends to which the contractors have gone to minister to the general comfort and well-being of their employees is without precedent in American contracting annals. In a sense, the Hoover Dam project is not only a construction job but also a sociological venture.

A tent colony, served with water from tank cars, housed the workmen who founded Boulder City. These men quickly put together forty 8-man frame houses, built on skids to facilitate their movement later on. A brigade of carpenters, plumbers, and other artisans occupied these houses as soon as they were ready, and proceeded to erect the contractors' permanent camp—the most extensive array of buildings ever assembled in connection with a construction job in this country. The contract provides that 80 per cent of the contractors' employees must live in Boulder City; and, in laying out the town, the Government allotted certain portions to the contractors for building purposes. A rental of $5,000 a month is paid for the use of this land.

Eight dormitories for single men, each two stories high and capable of housing 172 persons, have been built. They are in the form of a letter H, with shower baths and toilets in the central section. A large, screened porch runs the length of the building on each floor. A unique feature is the provision of individual rooms, 7x10½ feet, for each man. A large building houses the general offices of Six Companies Incorporated, in which are grouped the executive, accounting, purchasing, and engineering staffs. Adjacent to it is a dormitory for unmarried office employees, with one end made into a clubroom. Close to the structures previously mentioned are the mess hall, with a seating capacity of 1,000, and a modern laundry. A separate dormitory is provided for the 75 mess-hall employees. Another large structure serves as a recreation center. It contains billiard tables, a soda fountain, a barber shop, a news stand, and other features. A well-defined recreational and sports program is carried on under the direction of Frank Moran, former contender for the heavyweight boxing championship of

**A vast amount of scaling and other work involving "Jackhamer" drills was done on the canyon walls during early operations. At the right is one of the eighteen I-R portable compressors used before stationary air-producing plants were installed.**

the world. A commissary, which is in reality a department store, occupies a large building. Fixtures, built especially for the purpose, and furnishings of the most modern type make attractive the interior, which is divided into sections which retail every conceivable item of merchandise at prices which compare favorably with those asked for similar classes of goods in Los Angeles. All major buildings are equipped with water-washed air-conditioning plants and air-distribution systems which make it possible to cool or heat them as desired. Four central heating plants will be utilized in winter months. Electric water coolers are installed in all principal buildings.

Individual cottages, each placed on a 50-foot lot, are provided for renting to married employees and their families. Each cottage is supplied with electricity for lighting, flasks of high-pressure gas for cooking, and fuel-oil stoves for heating. It might be noted here that Boulder City is as near a smokeless town as exists. Up to February 1 of this year 396 such cottages had been built, consisting of 260 two-room and 136 three-room units. There are also eleven 5- and 6-room houses for construction officials and two larger residences for Superintendent Crowe and company executives.

In addition to these buildings are suitably situated warehouses, a garage, machine shop, etc. The machine shop, a steel frame structure, is completely equipped to handle repairs on machinery of all classes used on the job, ranging in size up to locomotives and huge power shovels. Compressed air is furnished by an Ingersoll-Rand Type 20 compressor of 316 cubic feet per minute piston displacement. Included in the shop equipment are several I-R air hoists. There is also an air-operated forging hammer.

A modern hospital of brick construction and containing $30,000 worth of equipment is designed to become a permanent feature, as it will probably be taken over by the Government upon completion of the contract. It offers every service that can be obtained in the average large city hospital. Twenty beds are now provided, with plans underway to add ten more.

The building program was in charge of the Boulder City Company, a subsidiary organized to direct feeding, housing, and transporting operations. V. G. Evans serves as its manager. George de Colmesmil of San Francisco was the architect. With a few exceptions the buildings have frames of wood, with stucco on the outside and wall board on the inside. Roofs are of asbestos shingles.

The high standard of food served has occasioned comment by all who have visited the job. The mess hall is undoubtedly the finest of its kind ever built for similar purposes. Two large dining rooms, equipped with facilities that savor more of a high-class restaurant than of a contractor's camp, are arranged on either side of a commodious kitchen that is outfitted in the most up-to-date manner. Ingenious machines to aid and improve the preparation of meals abound. Ranges fired by electricity, oil, and gas are available for cooking. Refrigerating rooms are provided for the preservation of meats, vegetables, and other foodstuffs. Incoming supplies are delivered direct from railroad cars or trucks by monorail.

The mess hall is operated under contract by the Anderson Brothers Supply Company, an organization of vast experience in shipyard and movie-location feeding. The finest foods obtainable are served, and there is a wide choice at each meal. Milk—500 gallons of it a day—comes by truck 80 miles from a 160-acre farm which was purchased especially as a source of supply. Last Thanksgiving the dam workers consumed, among other items, 2,400 pounds of turkey, 300 gallons of oyster soup, 300 pounds of cranberries, 760 pies, half a ton of plum pudding, and a quarter ton of candy at the principal meal. Everybody, from the highest executive to the lowliest laborer, eats the same food in the same dining room. Meals are served at six separate times during the 24 hours to accommodate the men on the various shifts. Workers going down to the canyon carry with

A battery of I-R portable compressors supplying air for beginning one of the four diversion tunnels. The insert illustrates how these machines were floated down the river to points of work before roads were built.

them a lunch which they select themselves in cafeteria style. An auxiliary camp is maintained at Cape Horn, two miles above the dam site on the river's edge. Six dormitories, each accommodating 80 men, a mess hall, commissary, and recreation room are provided there. They are similar in construction to the buildings at Boulder City.

The building program which has been described extended over a period of many months—in fact, is still going on. Principal structures were, however, erected with great speed, and by the end of the summer accommodations were available for 2,000 men. But it should be remembered, that while the camp in its finished state will eclipse anything of its kind previously built, it lacked many things during the early months of the construction period. With the thermometer registering more than 100° in the shade for days at a time, living conditions were necessarily hard. The chief handicap was the absence, through no fault of the contractors, of an adequate water supply. Consequently, a bath was a real luxury. Then, too, the cooling system for the buildings had not yet been placed in operation. Despite these and other shortcomings, which affected alike the highest and the lowest, the spirit of the great majority of the men was admirable. The fine coöperation which was shown by the labor force during the pioneer stage of activities was one of the primary reasons that the contractors were able to "get the jump" on the job in hand. Many a disposition which was sorely tried was helped over the rough spots by the knowledge that Six Companies Incorporated was doing all that was humanly possible to provide a maximum of comfort and convenience at the earliest possible moment.

Even while housing activities were in their infancy, the contractors plunged into the task of opening up operations in the canyon. This involved many things. The whole method of attack had to be decided upon, and an organization set up to handle each of the countless phases in its turn and to coördinate all of them. Surveys had to be run and plans drawn. Equipment of every conceivable kind, running from nails to trucks, had to be purchased and got on the job as quickly as practicable.

Of immediate concern was the matter of making the river bottom accessible to men and machinery. A 2-mile highway had to be hewn from virtually solid rock to connect the end of the Government road with the water level at a point near the site of the diversion tunnel outlets on the Nevada side, about half a mile below the dam site. At the upper end of the canyon, Hemenway Wash provided a natural approach to the river at a point some two miles above the dam site. From there a road, again mostly through rock, had to be constructed downstream along the base of the cliff to give access to the sites of the upper portals of the diversion tunnels on the Nevada side. Some 30 miles of power lines had to be run from the substation to various points in the canyon to serve electric shovels, electric-driven compressors, pumps, tunneling machinery, etc., and to furnish current for lights within the tunnels and outside flood lights for night work. A number of suspension foot bridges had to be thrown across

**S-49 "Jackhamers" excavating at the base of a cliff in the canyon for a bridge anchorage. A group of "cherry pickers", carrying on their hazardous task of scaling a canyon wall, is shown in the insert.**

the river to provide means of reaching the Arizona side with men and materials.

Location surveys had to be run and work started on some 21 miles of standard-gage railroad to connect the Government line with various portions of the work. The line takes off of the Government railroad to the dam site about half way from Boulder City and runs by a winding route, to secure grade, to a point in Hemenway Wash designated as Junction City. Here, at an elevation of 1,015 feet—75 feet above the high-water line of the reservoir that will be created by 1936—will be located the gravel screening and washing plant and the stock pile of aggregates for the concrete that will go into the dam. From Junction City a branch extends upstream to the Arizona gravel deposits and another branch downstream to reach the dam site at an elevation of 720 feet. The last-named section will deliver concrete for pouring the first stages of the dam.

When concreting operations are at their height, this railroad will carry a volume of traffic heavier than that on any main line in the country. Just for transporting the 5,000,000 cubic yards of gravel that will go into the concrete for the dam it will require the equivalent of a 1,300-mile train of freight cars each loaded with 100,000 pounds. Sixteen miles of grading on this railroad work was contracted to the John Phillips Company, and the track laying was contracted to Shannahan Brothers. Six Companies forces undertook the building of the 5-mile section from Junction City to the dam site. Two miles of this work was in the solid rock of the canyon wall, and much of it in tunnels or half tunnels.

Another phase of the work which was started early was the scaling of loose or projecting material from the canyon walls. This cleaning off was done not only to clear sites for the numerous tunnel and adit openings but also as a measure of safety to protect workmen in the canyon bottom.

Compressed air played a vital part in essentially all these pioneer aspects of the construction. At one time eighteen Ingersoll-Rand Type 20 and Type XL portable compressors were in service by Six Companies Incorporated. Before the road reached the canyon bottom, barges were employed to float some of these machines downstream from a landing at Hemenway Wash to suitable locations. Incidentally, water transportation has solved the problem of reaching the many points of work which are inaccessible by roads. The contractors early put into service a fleet of motorboats which continue to be of great assistance and which likely will be used to a considerable extent throughout the life of the contract.

"Jackhamers" were indispensable tools in aiding the highway, railroad, and scaling operations, and later on in facing openings for the four diversion tunnels. The drill selected for all these purposes was the Ingersoll-Rand Type S-49.

With the varied activities enumerated in progress and many other divisions of the work being started, the intensity of operations that prevailed can hardly be comprehended. During September, 1931, approximately 50 tunnels were in various stages of completion. Meanwhile, much Government work, which will be described in a later article, was also being vigorously prosecuted.

**Good food and recreation for employees are considered important by the contractors. From top to bottom the pictures show: the recreation building and, at the right, the department store; a corner in the recreation building; a portion of the spick-and-span kitchen; one side of the 1,000-seat dining hall.**

Three of the larger structures erected by the Government in Boulder City. The dormitory and guest house is shown above, the administration building in the center, and the post office below.

# Construction of the Hoover Dam

*Within a Year's Time the Government has Reared a Modern City in the Desert at a Cost of $1,600,000*

C. H. VIVIAN

BOULDER City, Nev., is a municipal by-product of the Hoover Dam. Sired by Uncle Sam, and growing up under his watchful eyes, it is unique among American cities. A year ago the ground it occupies was desert waste. Today it is a community of more than 4,000 souls and some 600 buildings. It has paved streets, water, sewers, electric lights, and telephones. Soon it will have lawns, flower gardens, shade trees, and a park.

Even the Government is in doubt as to its ultimate size and prosperity; but its immediate future is assured by virtue of the clause in the Hoover Dam contract which requires 80 per cent of the employees of Six Companies Incorporated to live there during the construction period. After the dam is completed, nobody knows just what will happen. A certain number of Bureau of Reclamation men will be stationed there, and probably the employees of the power company that takes over the operation of the huge generating stations will also make it their home. As regards other permanent residents, the Government is only speculating; but it has reason to believe that the vast amount of publicity given the dam will draw homeseekers to the region and that the completed project will prove a magnet for tourists. At all events, the development plan visions an enduring town of 4,000 or 5,000 persons.

Boulder City is not the creature of whims or theories. It was born of sheer necessity. Bureau of Reclamation officials recognized that the Hoover Dam could never be reared in the midst of a desert unless precautions were taken to insure adequate care of the workmen. The bureau knew this because its own engineers had carried on in the blistering heat of many summers gathering the preliminary data required to locate and plan the structure now in the making. Their home was a cluster of tents which afforded little bodily or mental comfort. Fortunately, there were relatively few of them, and they managed to endure discomforts and to ward off pestilence. It would be out of the question, however, to house 3,000 men in this manner for six or seven years.

It would be a bit ironical, the Government thought, if the climatic restrictions proved too severe for American labor to carry through the largest single construction project it had ever attempted. And there were plenty of persons who predicted this would come to pass. Some prophesied that Orientals would finally have to be imported to cope with the melting temperatures that prevail in Black Canyon for four months out of the year.

To make certain that nothing of this kind would happen, Uncle Sam decided to extend a kindly but firm paternal helping hand to the contractors and to set up a construction town under his control to insure for the workers a high standard of living and a maximum of comfort and general well-being. The wisdom of this course is shown by the marked effect it had on the attitude of bonding companies. Two of these concerns sent an investigator to the dam site before the contract was awarded. This emissary sizzled in the canyon one day and then took back his report. It was enough. The companies let it be known that they had no interest in underwriting the performance bond required of the contractors, despite the fact that the premium promised to exceed $1,000,000. But when they learned the Government's plans for improving living conditions, they reconsidered the matter and ended by giving to the project a far lower rate than had been charged on many enterprises of smaller size.

The Government set aside $2,000,000 for

**Boulder City, as it appears from a hill to the north. In the foreground are residences for Government employees. At the extreme left are the administration building and the dormitory. The group of small buildings at the left center are residences of employees of Six Companies Incorporated. In line with them, at the right, are the contractors' dormitories.**

the construction of Boulder City. Of this sum approximately $1,600,000 had been spent or contracted for up to March 1, 1932. Great care was taken in selecting a site. A town in the canyon itself was out of the question. There wasn't room for one, and the terrific heat during a third of the year forbade it. Dr. Elwood Mead, Commissioner of the Bureau of Reclamation, summed up the situation in the following words:

"In the summer the wind that sweeps over the gorge from the desert feels like a blast from a furnace. At the rim of the canyon their is neither soil, grass, nor trees. The sun beats down upon the broken surface of lava rocks. At midday they cannot be touched by the bare hand. It is bad enough as a place to work. It is no place for a boarding house or a sleeping porch. Comfortable living conditions had to be found elsewhere, and these are found on the summit of the divide, seven miles from the dam site. Here there is fertile soil; here winds have an unimpeded sweep from every direction; here there is also an inspiring view of deserts and lonesome gorges and lofty mountains. When the dam is completed and a marvelous lake fills the foreground, the view from Boulder City will be so inspiring and wonderful as to be worth traveling around the world to see." The city site has an average temperature lower than that at any of the others considered and several degrees lower than the temperature at the dam site. It is 2,500 feet above sea level, 1,700 feet above the river, and 1,250 feet higher than the top of the future dam.

During the life of the Hoover Dam contract, Boulder City will be a municipality with two distinct sections. It was told in a previous article in this series how Six Companies Incorporated has spent upwards of $800,000 for buildings that will be razed when the dam is finished. In contrast to these temporary buildings, the Government is providing the permanent structures of the town. These include, aside from the site itself, the various utilities such as water and sewer systems and street lights, also sidewalks, paving, parks, public buildings, and permanent dwellings for its employees.

It was originally intended that Boulder City should be built largely in advance of the coming of the dam workers, who were to move into a spick-and-span town affording all conveniences. This plan was abruptly ended by the decision to speed up the dam contract to help alleviate the business depression. Thus it happened that when the contractors moved on to the job in March, 1931, the city was still in the blueprint stage. As a result, many hardships had to be borne during the first summer that adherence to the first plan would have averted. Happily, the next hot-weather period will find Boulder City better prepared for it.

The town site includes approximately 300 acres, and is laid out roughly in fan shape with the point at the north, where a ridge separates it from Hemenway Wash. The northern portion is allotted to Government buildings. Groups of residences for Bureau of Reclamation employees are arranged along the lower slopes of the ridge on either side of the fan point. Between these two groups are the administration building and the dormitory. Just below these, to the south, is a 5-acre park at the apex of an angle formed by the meeting of two principal streets. These thoroughfares diverge to the far corners of the city at the south and form the chief arteries of travel, with most of the town site between them.

The ground has a fairly uniform slope of about 3° from north to south. The central portion of the city will be devoted to business and commercial uses; the southern section is set aside for residences. Considerable areas in this district is now occupied by temporary frame houses for employees of Six Companies Incorporated. Most of the contractors' larger structures, such as dormitories, store, recreation building, etc., are on the western edge of the town, where they do not impede development of the central part of the city.

Boulder City was laid out by S. R. DeBoer, a well-known city planner, and is designed to be a model town. In the business section are being provided special plazas for the parking of automobiles, and no street parking will be allowed. Alleys in the commercial zone are laid out 50 feet wide to permit the loading and unloading of trucks in the rear of stores and thus to lessen their use of street space. Through automobile traffic will travel on highways which are separate and distinct from the business and residential streets. Streets will be graduated in width according to their intended usage. Through highways and business streets will be 92 feet wide, and residential streets 60 feet wide.

In the districts which are expected to have

**office buildings, etc. The business section of the city will be developed near the center, where the arcaded post-office building may be seen. The insert at the left shows the site of Boulder City as it looked early in 1931. The insert at the right shows the tent camp, near Boulder City, which was the home of the Government engineers until a few weeks ago.**

the greatest density of population, the blocks are laid out 900 feet long and 260 feet wide. In their design, provision has been made for interior plazas which will contain small parks. These can be equipped with playgrounds for small children and with croquet lawns and horseshoe pitching courts for elderly persons. The older boys and girls and the young men and women of the future Boulder City will find recreational facilities in the community park in the form of football, baseball, tennis, and other sports which are being planned for their amusement.

The Government's program of building, which is now well advanced, includes a $60,000 administration building, a $45,000 municipal building and post office, a $33,000 dormitory for unmarried workers and visitors, a garage and fire station, a schoolhouse, 100 residences of from three to seven rooms each for married employees, and several community garages. Durable materials such as brick, tile, and stucco are being used in the buildings, all of which follow the Spanish type of architecture. The larger structures have furnaces, and the administration building is equipped with an air-conditioning system. Landscaping, and the planting of trees and lawns, have been started. When completed, these various items will have cost some $600,000.

During the past autumn and winter, the New Mexico Construction Company of Albuquerque, N. M., has been carrying out a $300,000 contract covering the installation of water distribution and sanitary sewerage systems; the grading of streets, alleys, and automobile parking spaces; the laying of concrete curbs and gutters and of concrete and gravel sidewalks; and the paving of certain streets and the surfacing of others. This work involved, among other operations, more than 23 miles of trenching for pipe lines, 151,000 cubic yards of excavating, and the laying of 90,000 square yards of asphaltic-concrete pavement.

One of the major tasks in connection with the building of Boulder City was the provision of an adequate supply of pure and palatable water. This was done at a cost of approximately $500,000 by installing equipment to pump Colorado River water more than six miles, elevate it nearly 2,000 feet, remove the discoloring sediment it carries, soften it, and treat it chemically to render it thoroughly safe for human consumption. A supply of greater initial purity could have been obtained by driving wells to tap the artesian reservoir beneath Las Vegas and the adjoining area, but cost and economic studies indicated that the river water could be developed at less expense.

The intake is just below the outlets of the diversion tunnels, roughly half a mile below the dam site. Three centrifugal pumps elevate the water about 100 feet to a 200,000-gallon pre-sedimentation clarifier. The pumps are mounted on a car which may be lowered or raised on rails of 47° slope, making it possible always to take water from a zone 4 feet beneath the surface, regardless of the stage of the river flow. Most of the sediment settles out readily, and the retaining of the water up to three hours in the pre-sedimentation basin serves to remove about 95 per cent of the solid matter. A Dorr traction clarifier mechanism, having a capacity of 62 tons of dry solids a day, removes the sludge which settles out.

Two lifts or stages of pumping are required to raise the water from the clarifier to the treatment plant at Boulder City. The pumping equipment at both stations consists of three 4-stage, 450-gallon-per-minute, 1,200-foot-head centrifugal pumps. The lower units force the water through 20,000 feet of 10- and 12-inch steel pipe to a booster station almost 1,100 feet higher. The second bank of pumps, which takes its supply from a surge tank 90 feet high, pushes the water through 14,500 feet of pipe and raises it 850 feet to a 100,000-gallon tank in Boulder City. Extensive treatment and filtration equipment is installed there to soften the water and further to clarify it. Following filtration, the water is chlorinated and pumped to a 2,000,000-gallon storage tank on a high knoll just north of the town. Three 500-gallon-per-minute, 170-foot-head pumps serve to give it this final lift of 150 feet. All pipe in the system is buried approximately 3 feet deep.

The system was designed by Burton Lowther, of Denver, consulting hydraulic and sanitary engineer to the Bureau of Reclamation. It was built under a number of contracts covering its various phases.

River water, untreated except for pre-sedimentation and chlorination, was delivered to the storage tank for the first time on September 9, 1931, thereby providing an emergency supply for the city. Up to that date, Six Companies Incorporated had been hauling water in tank cars from Las

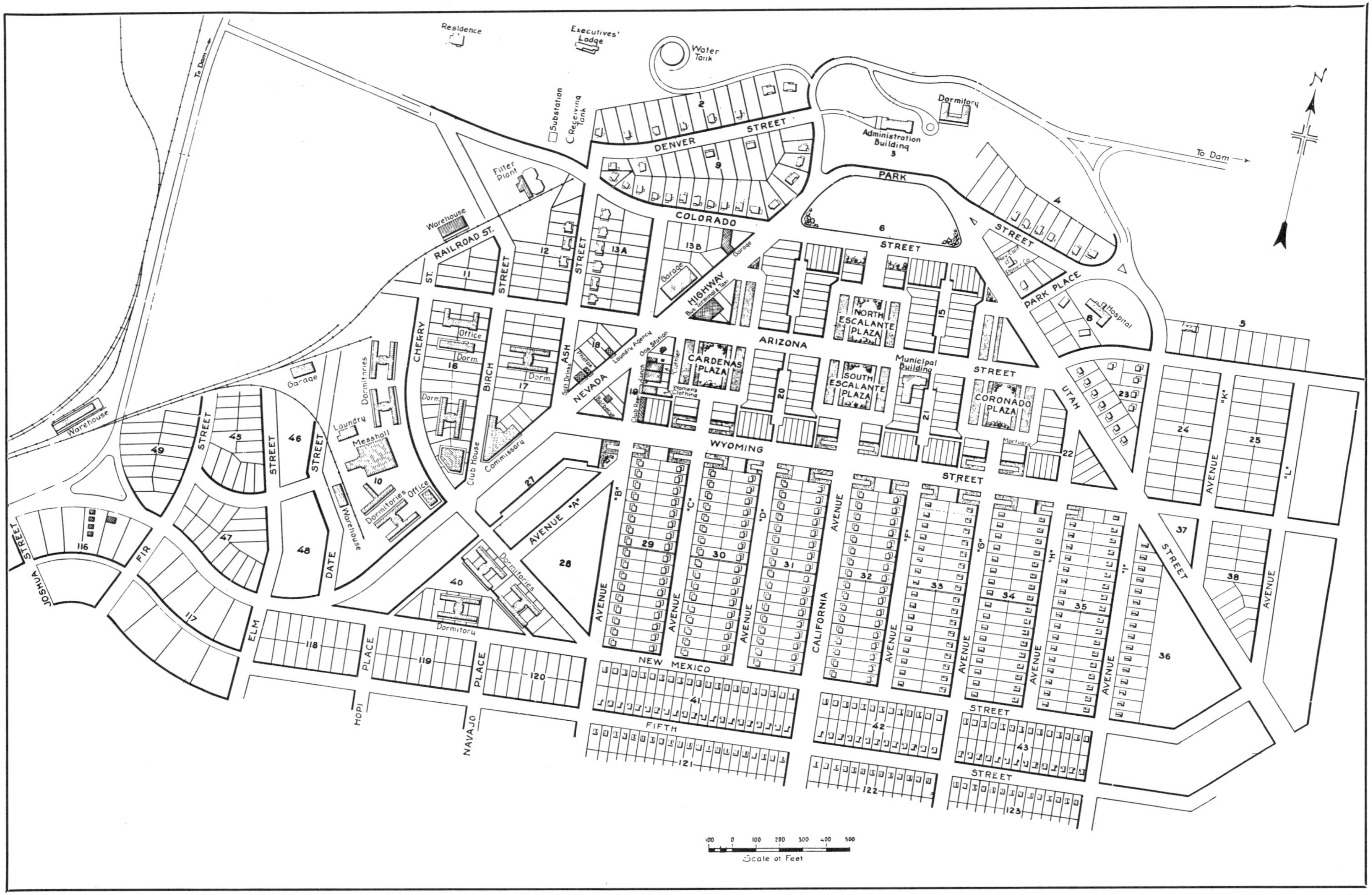


U. S. Bureau of Reclamation

**Plan of Boulder City which shows the location of buildings erected or in course of erection. Most of the structures at the left and in the lower section of the city are used by Six Companies Incorporated and their employees.**

Top view at the left is of the intermediate pumping station. Below is the treatment and filtration plant in Boulder City that is a model of its kind.

At the right the top view shows the movable pumps at the river and the pre-sedimentation and pumping structures above. Below is the pre-sedimentation basin.

Essential structures in the Boulder City water-supply system.

Vegas. The daily consumption during the torrid summer months was as much as 50,000 gallons, and cost from half a cent to three-quarters of a cent a gallon. The city distribution pipes and the treatment plant were not installed until months later, so that the water system was not functioning in its completed form until February, 1932. A sewage-disposal plant is being constructed about a quarter of a mile from the city limits.

An area of 110 square miles, of which Boulder City is a part, was withdrawn from public entry in 1921 and, except for a few pieces of patented ground, is entirely under Government control. While never formally declared a reservation by Congress, it was established under Nevada statutes. Vistors approaching Boulder City must halt at a sentry station, and passes must be procured before proceeding further. Contractors' workman are hired in Las Vegas and given credentials to admit them to the reservation. Sight-seers bent upon having a look at the construction drama in the canyon are shown every courtesy, but they are not allowed in the working zone. The nearest vantage point for them is Lookout Point, a rocky crag nearly 800 feet directly above the dam site on the Nevada side, where a place of observation has been established.

The Government intends to retain jurisdiction over all the land in Boulder City during the construction period in order that it may have full control of activities there. No business or professional enterprise can be established in the town except under Government permit. Land for business or residential purposes cannot be bought, it can only be leased; and structures built thereon must be approved from the plans before construction can start. Leases run for a maximum of ten years, being limited because of the uncertainty of Boulder City's status after the dam is completed. The future disposition of such leases will be determined upon their expiration. They may be extended, or the land may be sold.

Some Government residences and, above them, the 2,000,000-gallon water-distribution tank on a hill above Boulder City.

Establishment of these stringent regulations was deemed advisable to prevent a wholesale influx of people, with consequent unemployment, ruinous business competition, and widespread infractions of laws. Even before the dam contract had been awarded, the Bureau of Reclamation was deluged with inquiries from persons in all parts of the country who had read of the great undertaking and who desired to engage in business in the projected town. Most of these people had no conception of the climatic conditions, of the actual number of men that would be employed, nor of the business opportunities that would be available.

In March, 1931, Secretary of the Interior Wilbur appointed Louis C. Cramton, former member of Congress from Michigan, to assume charge of the appraisement of lands in Boulder City and of the making of lease concessions. Mr. Cramton formulated regulations governing the granting of permits for conducting business and for leasing ground, and these he published in a pamphlet together with a general description of conditions that would be met with in the construction town This information was issued on May 18, by which time more than 3,000 letters of inquiry had been received and placed on file by Jesse W. Myer, chief of the mail and files section of the Bureau of Reclamation, who had been detailed to Las Vegas to classify them for handling. Additional mail inquiries were coming in at the rate of fifteen a day, and as many more persons were calling at the Las Vegas office. It having been decided that only formal requests for permits would be considered, a pamphlet and an application blank were sent to each inquirer. A $10 fee to show good faith was required with each application, the sum to be returned if no permit were granted. The informal inquiries continued in volume and totaled more than 4,000 up to October 1, 1931, but on that date only 320 persons had returned the official blanks accompanied by the fee.

The system adopted effectively curbed a mushroom growth and undoubtedly averted confusion, disappointments, and hardships. Said Mr. Cramton at the time the regulations were drawn: "There is no doubt that, if the Government desired, it could create in Boulder City in the next year one of the most spectacular boom towns in recent history. If we were to reply to inquirers without discouragement and without limitations, simply setting aside the necessary lots for business and residential purposes, I have no doubt a thousand or more persons would sell out what they have at home and go to Boulder City expecting to make their fortunes there. Ruin would inevitably follow any such movement, for the business possibilities are limited. Certainly the Bureau of Reclamation does not desire to have any part in such wholesale business disaster."

**Sims Ely, city manager of Boulder City.**

Formal applications for business permits were received from persons in 36 states and covered more than 60 types of business, ranging in alphabetical order from automobile sales to welding. Had all who applied set up establishments, there would have been in Boulder City 31 drug stores, 21 indoor recreation rooms, 16 barber shops and beauty parlors, 14 restaurants, 14 filling stations, 12 soft-drink shops, and innumerable stores of other kinds. To guard against such over-commercialization, the Government set up four classifications to guide the granting of permits:

1. Exclusive—Public utilities and similar operations. Only one permit for each classification.
2. Limited—Mercantile stores, such as groceries and markets. At least two competing permits in each line.
3. Special—Banks, motor lines to outside points, etc. The number of permits to be governed by the prevailing conditions.
4. Personal—Professional services such as doctors, lawyers, and dentists. Permits to be granted subject to authorization in the states where the respective applicants reside.

The field for business enterprises is narrowed through the fact that Six Companies Incorporated maintain various retail establishments essential to the needs of their employees, including a dining hall, department store, recreation hall, laundry, and barber shop. A business permit carries with it the right to lease ground for business purposes. The average annual rental for a business lot 40x120 feet in $275. Any person of good character may lease ground for residential uses at an average annual rental of $120 for a lot having a 50-foot frontage.

Up to March 1 of this year 114 business permits had actually been granted, and 26 retail and wholesale establishments had been set up. Buildings erected or in course of erection for these purposes totaled eighteen. Twelve permits to lease residential lots had been issued, and two homes had been built or were under construction.

Sims Ely, of Arizona, heads the administrative staff of Boulder City as city manager. He was appointed by Secretary Wilbur and reports to Walker R. Young, construction engineer in charge of the project for the Bureau of Reclamation. He is assisted in governing the town by an advisory board of three men, of whom two represent the Government and one represents the contractors.

The policing of Boulder City, as well as of the entire reservation in which it is located, is in the hands of deputy United States Marshals. A force of nine men is headed by a chief ranger. G. E. Bodell is chief of police of Boulder City. Laws prohibiting gambling, the sale of liquor and narcotics, and other practices which the Government deems injurious to the workers and to the orderly progress of the work in the canyon are rigidly enforced.

**Left—The first office of the United States Marshal. Center—Chief of Police Bodell of Boulder City flanged by two members of his force. Right—The gateway to the reservation where all vistors must secure passes.**

**One of the drill carriages, with drills in place, shown outside a tunnel portal. The man at the right, in the foreground, is wearing a metal hat of the type adopted for the protection of workmen who are exposed to the hazards of falling rocks.**

# Mammoth Drill Carriages Speed Hoover Dam Tunnel Work

ALLEN S. PARK

THE accompanying illustrations show the rock-drill carriage which is materially aiding Six Companies Incorporated in driving the huge tunnels that will carry the Colorado River through the solid rock walls of Black Canyon while the Hoover Dam is being built.

This type of carriage, which is believed to be the largest ever constructed, makes it possible to mass 24 to 30 Ingersoll-Rand N-75 drifter drills in a simultaneous attack against the tunnel breast, and is perhaps the most important factor in enabling the contractors to carry on tunneling operations at a much faster rate than was thought possible prior to the beginning of the work.

The carriage or jumbo, as it is familiarly known on the job, is the creation of Bernard Williams, general foreman of work in the canyon. He designed and directed construction of the first unit, which proved so effective that similar devices were built and adopted for all the enlargement work.

Each of the four diversion tunnels will be 56 feet in diameter and approximately 4,000 feet long. Considering their length, they are of record size. The Rove Tunnel in France, completed by the French Government in 1927, is 78 feet 6 inches wide and 54 feet 4 inches high, but of short length. No other bore through rock has ever equaled the overall proportions of the Hoover Dam tunnels.

It is obvious that the advancement of such enormous faces presented perplexing problems. The operations required are on such a vast scale as to almost fall in the category of quarrying rather than of tunneling. Six Companies Incorporated considered many possible systems of procedure before beginning actual work. The plan suggested by Mr. Williams was the first one tried; and its success saved much time and money which might otherwise have been expended in experimentation.

The general plan of tunnel advancement was to drive 12x12-foot pioneer headings at the top of the cross section of the 56-foot area ultimately to be taken out. Enlargement was started, however, long before these smaller headings had been completed. In fact, the drill carriage was first tried out at the lower portal of Tunnel No. 4 where the pioneer tunnel had not yet been started. Even there, though, a heading 12 feet high and extending to the arch lines of the completed size was kept one round in advance of the bench. Thus, the procedure was essentially the same as though the pioneer bore had already been driven. As they are now being excavated, the enlargements are carried to their full width but only 40 feet high. This leaves 16 feet in the bottom or invert to be taken out later. With a 12-foot heading excavated at the top, there remains a bench 30 feet high, 56 feet across at its widest line, and 50 feet in average width.

"We wanted some method of driving the tunnels without using ring drilling and without putting in down holes," said Mr. Williams in discussing how they approached the prob-

**Two views of a "jumbo" at a tunnel heading. At right drillers are pointing their machines preparatory to starting a round. The other picture shows a carriage, mounted on its 5-ton truck, backed into position for drilling one side of the 56x30-foot bench.**

lem of developing the drill carriages. "Since we had to carry the tunnels at least 40 feet high to allow working headroom for the Marion Type 490 electric shovels to be used in excavating, a 12-foot heading and a 30-foot bench were determined upon as the best plan of attack. A method then had to be devised to drill flat holes in the bench. A drill carriage of some sort seemed to offer the greatest possibilities of filling the requirements. As we intended mucking into trucks, a truck-mounted carriage was the logical solution. It was essential that the drill carriage and the shovel be able to pass in the tunnel, so we decided upon a carriage wide enought to drill half the bench at one set-up."

As can be seen in the pictures, the drill carriage has a steel skeleton which supports wood platforms at four levels. Two of these platforms extend the full length of the carriage and provide working sta ions for the drilling crews. The other two platforms are of shorter length and serve as drill-steel racks. They are partioned so that each machine has its set of steel always handy.

The drills are supported on transverse pipe bars secured to the outside of the frame uprights at five levels. Extension arms at the ends of these bars permit of setting up drills for driving holes at suitable angles into the side walls to break the rock approximately to the curve line of the finished excavation desired. Four of the five lines of drills are operated from the two platforms, and the lowest one is operated from the ground.

The carriage is piped for air and water immediately below the drills. A 6-inch air pipe and a 2-inch water pipe are run along the floor of the tunnel at one side up to within 100 feet of the bench. These services are extended to the carriage through the medium of three 2-inch hose lines, two of them for air and one for water. The lines connecting the drills with these sources of supply are 5 feet long. Thus each machine can be moved laterally 4 feet either way from its central position, which is ample to meet all drilling conditions. An apron of sheet stee' built above the drills protects the operators from falling rock. The steel uprights of the frame at the four corners of the carriage reach nearly to the ground. This perimts blocking the carriage solidly in drilling position with screw jacks. As a result, vibration has never proved bothersome, even with 24 drills running.

To set up the carriage for work it is necessary only to back the truck into position, connect the water and air lines, block up the corner posts, and point the machines. The average time required for these operations is twenty minutes, but they have been performed in as few as ten minutes.

After half of the bench has been drilled from one setting of the carriage, the truck is pulled away and backed into position for drilling the other side. When this second half has been drilled, the truck is moved out of the danger zone and the entire round is blasted at one time. After the blast the power shovel is brought up to the face to load the muck into trucks. With the operations thus systematized, there is a minimum loss of time. Records kept for one week at the beginning of the enlargement work in all the tunnels show that the average drilling time was 4.38 hours, the average mucking time 6.95 hours, and slack time only 2.96 hours. Thus the total average time required for a complete round was 14.29 hours. The average advance for each round was 15.11 feet. These figures were considerably improved upon as the crews became better trained; and before the first tunnel was holed through the average advance per round was not only increased but at some faces three rounds were completed in 24 hours.

The success of the drill carriage on the enlargement work has prompted the engineers for Six Companies Incorporated to modify it and to use it for drilling the remaining bottom section of the large tunnels. Detailed information on this work is not yet available but it is the concensus of opinion that the drill jumbo can be utilized to very good advantage.

Looking out of the gaping maw of one of the huge diversion tunnels.

# Construction of the Hoover Dam

*Details of the Driving of the Four Huge Tunnels which will Divert the Colorado River Around the Dam Site*

**NORMAN S. GALLISON***

THE first major operation confronting Six Companies Incorporated in the building of the Hoover Dam was the driving of four great tunnels through solid rock. These will carry the Colorado River around the dam site while excavation for the dam foundation is underway, and thereafter until the massive concrete barrier has been partially erected. There are two of these diversion bores on each side of the river. As driven, they are circular in cross section and 56 feet in diameter. After being lined with 3 feet of concrete, they will have a finished section of 50 feet. Their combined length is 15,909 feet, and their individual lengths are: No. 1—4,300 feet; No. 2—3,879 feet; No. 3—3,560 feet; and No. 4—4,170 feet.

At the dam site, the river flows through a narrow box canyon whose walls rise sheer from the water's edge to a height of 800 to 1,000 feet. The tunnels enter these precipitous cliffs about 2,000 feet upstream from the axis of the dam site, follow roughly semicircular courses through the walls, and emerge approximately 2,000 feet downstream from the dam site. The beds of these passageways are a few feet lower than the normal river level. From intake to outlet, each bore has a fall of about 14 feet. The intake portals of the two tunnels on each side of the river are relatively close together but, owing to the conformation of the canyon walls, they are not truly parallel and are farther apart at their outlets.

After these tunnels have served their primary purpose of diverting the river's flow, they will not be abandoned. The two nearer the river will be closed off with concrete plugs, above and below the dam. In the lower plugs will be installed needle valves, which will serve, through the penstock pipes, as outlets for regulating the flow of the river below the dam while the reservoir is filling to the elevation of the intake towers. These tunnels will also house the pressure penstocks from the intake towers for a portion of their length. The outside set of tunnels will be bulkheaded at the upper portals and about midway in their courses, from which latter points steeply inclined tunnels driven to the surface will, when equipped with suitable structures at their tops, act as spillways for the overflow waters.

The driving of the four tunnels involves the excavating of nearly 1,500,000 cubic yards of rock within the tunnel lines. A further vast amount of material was removed in gaining access to the various points of work. Operations were started in May, 1931, and at the present time it seems reasonably certain that they will be completed during the present month, so that the total elapsed time required will have been about a year. This is considerably under the best estimates made prior to the beginning of work. Operations have been carried on 24 hours a day, in three 8-hour shifts. As many as 1,500 men have been engaged in the tunneling activities, and the average number employed has been 1,200.

*Public and Press Relations Division, Six Companies Incorporated.

**Above—Bench drill carriage and full crew.**

**Above—Drill carriage at work in tunnel.**

**Right—Lower tunnel portals during flash flood.**

**Drilling crew in 12x12 top heading.**

**The boom swings and 3½ cubic yards of muck is loaded.**

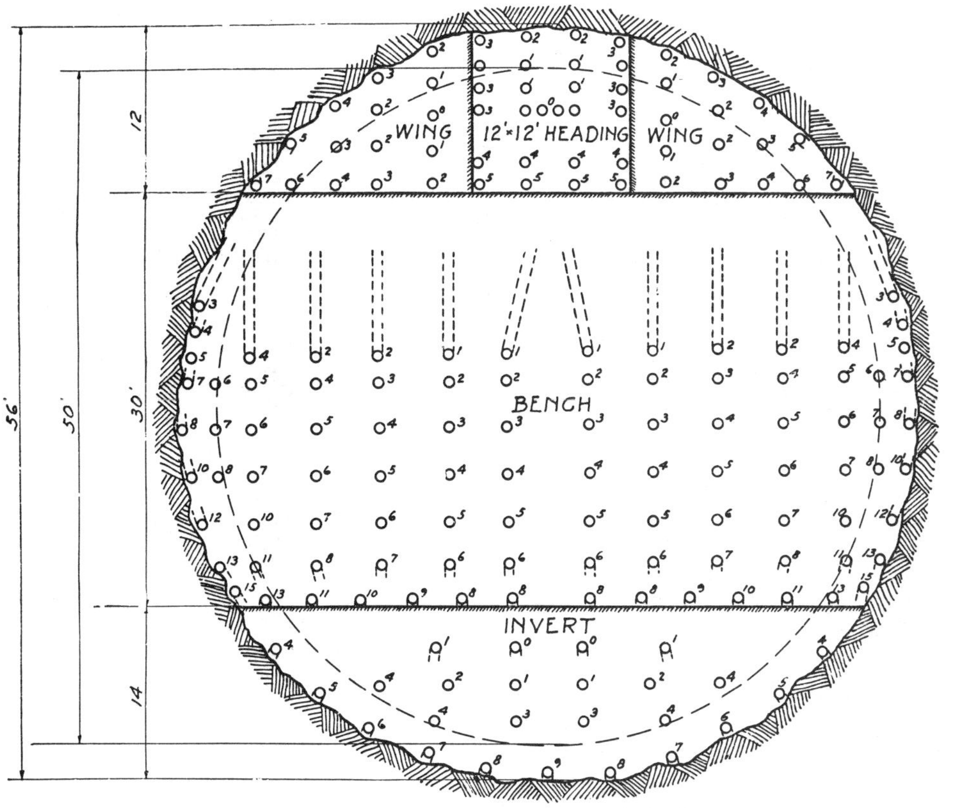


**Holes were located as shown above and fired in the various sections in the order indicated.**

As soon as they took hold of their contract, Six Companies made preparations to begin tunnel driving, with the result that drills were biting away at the rock several weeks before a roadway had been constructed into the bottom of the canyon. About two miles above the dam site, the canyon broadens out into the reservoir area. There, at the foot of Hemenway Wash, access to the river level by trucks could be had over a rough desert road. Barges were constructed at this point; Ingersoll-Rand Type 20 and Type XL portable air compressors and drilling tools were loaded aboard; and the first expedition set forth. The only spot in the canyon proper where a foothold could be obtained was on the Arizona side of the river, near the center of the future dam. At this point was a small talus slope formed by rocks which had fallen from the cliffs above. Workman on the barge, which was lowered downstream by cable, were able to moor their craft and unload compressors, a portable blacksmith shop, drills, and other equipment among the rocks.

The plan of attack adopted by Six Companies incorporated was to drive adits into the canyon walls on either side of the river to intersect the lines of the diversion tunnels midway between portals. The first hole was drilled for the Arizona adit on May 12, 1931. Ingersoll-Rand Type N-75 drifter drills and Type S-49 "Jackhamers" were used. The muck was hand shoveled into 1-cubic-yard mine dump cars and deposited along the canyon wall to enlarge the bench for further buildings. A cable suspension footbridge was then thrown across the river to gain access to the Nevada side. There a shelf was blasted from the cliff, and operations were begun in the same manner as on the opposite shore. These adits were driven 10x8 feet in cross section. The Arizona entry was 850 feet long, and that on the Nevada side 630 feet long.

Electric power was available on June 25, 1931. Prior to that time two Ingersoll-Rand Type XRE stationary compressors, each with a 1,302-cubic-foot piston displacement, were installed and connected to two 200-hp. Atlas diesel engines. These units furnished air for driving railroad tunnels and for roadbuilding. When power lines had been strung, the machines were converted to electric drive. About the same time a compressor plant near the Arizona adit was placed in operation, and this supplied air at 105 pounds pressure through 4-inch lines to the adit drilling crews. The plant consisted of two Ingersoll-Rand Class PRE units, each of 2,195 cubic feet per minute piston displacement. Later another compressor station was established near the lower portals of the tunnels on the Nevada side. It contained four Type XRE and two Class PRE machines having a combined piston displacement of 9,598 cubic feet per minute. All three of these compressor plants were interconnected by 6-inch lines; and air was distributed to the various points of use through a system of 4- and 6-inch mains having an aggregate length of some 30,000 linear feet.

BERNARD WILLIAMS
*Assistant Superintendent*

Ten-ton Baldwin-Westinghouse storage-battery locomotives and 3½-cubic-yard Western dump cars were brought in by barge and used to haul muck. When the lines of the diversion tunnels were reached, 12x12-foot top headings were opened off the adits in both directions. The purpose of driving the top headings in the main bores was twofold: to provide ventilation and convenient access for the enlarging operations which were to follow, and to furnish accurate information as to the character of the rock to be encountered throughout the length of the tunnels. By driving the adits, eight faces were opened up for attack in addition to those at the portals. Actually, the bulk of the footage in the top headings was driven through the adits. As soon as access could be had by cutting a road down the steep canyon walls, top headings were opened up at the lower portals of the two tunnels nearer the river and driven upstream toward the adits.

When electric power became available, top headings were drilled from a rig consisting of two horizontal bars or vertical columns mounted on a drill carriage running on the narrow-gauge rails used for mucking. Thirty-two holes, 10 feet deep, were drilled to a round, and the average depth broken per round was 8.3 feet. Mucking was done by Conway mucking machines. Crews working from the adits alternated as to headings, drilling one face while mucking operations were underway at the other. Although sixteen possible faces were available, actual conditions prevailing on the job were such that twelve faces was the most ever worked at one time. Progress in the top headings by months was as follows:

| | | | |
|---|---|---|---|
| June | 410 ft. | October | 4,147 ft. |
| July | 1,045 ft. | November | 3,025 ft. |
| August | 1,439 ft. | December | 1,191 ft. |
| September | 3,235 ft. | January | 120 ft. |

It will be noted that the total of 14,612 feet does not correspond with the combined length of the diversion tunnels. This discrepancy is due to the fact that the main bench operations were started on several of the portals before the top headings were completely driven through from the adits. Rock excavated from the top headings totaled 79,000 cubic yards. Much of this muck was dumped along the canyon walls to form roadways from the adits to the lower portals.

In gaining access to the lower portals of the diversion tunnels for the enlarging operations, considerable open-cut excavating, as well as scaling down of the canyon walls, was involved. The lower portal of Tunnel No. 4—the tunnel farthest from the river on the Arizona side—was the first prepared, and the initial round in the enlargement was fired on September 21, 1931. Since the top heading had not yet been holed through from the adit, a small 12-foot bench in the arch section of the tunnel was advanced one round, and then followed taking

**Above—Close-up of an invert drilling crew.**

**Left—Invert drill carriage and trimming jumbo behind it.**

off a bench 30 feet in height. This latter stage in the enlarging operations was referred to by the contractors as the "bench heading", as distinguished from the top heading discussed previously and the invert section which was removed at a later operation. Upon its completion, the opening was approximately 42 feet in height and horseshoe shaped in cross section.

After a short period of experimenting, the method adopted by Six Companies Incorporated for driving the bench heading was by means of a drill carriage or jumbo mounted on a truck chassis. An experimental rig built of timbers was first constructed, and proved to be successful. Later, several rigs of welded steel were built in the company shops and mounted on International truck chassis of long wheel base. These drill carriages were equipped with five horizontal bars, with six Ingersoll-Rand Type N-75 drills mounted on each bar. A more complete description of these drill carriages was given in a previous article of this series. One-half of the main 30-foot bench was drilled with one setting of the carriage. The truck was then moved forward and backed into position on the other half of the face while the side previously drilled was being loaded. Meantime, the wing sections on either side of the top heading were being drilled from two vertical bars. Drilling of the wings was also carried on during the mucking operations on the main bench.

Forty per-cent gelatine dynamite was used throughout. The powder was hauled by trucks from central magazines to smaller magazines near the points of use, and thence into the tunnels as needed. Primers were prepared by powder-fitters, working in isolated houses, and were carried into the tunnels in specially designed containers. As can be seen from the diagram on page 25, primers were set from "no delay" up to "15", making sixteen delays. A 440-volt circuit, with locked safety switches outside the tunnels, was used for detonating. Careful attention to wiring the leg wires to the buses resulted in remarkably few missed holes. The finally adopted method of placing the holes in the various stages of drilling, and the firing sequence, are shown in the diagram.

Electricity at several different voltages was required in the tunnels. The shovels used power at 2,300 volts. A 440-volt circuit was provided for blasting, and a separate lighting circuit of 120-volts was maintained. The tunnels were lighted throughout by reflectors hung along the walls up to within a few hundred feet of the face. Additional lighting was provided at the various faces by portable reflectors of 1,500-watt capacity. Water was pumped for use in drilling—in many cases directly from the river—through 2-inch pipe laid on the tunnel floor.

FLOYD HUNTINGTON
*Tunnel Superintendent*

Aside from the ingenuity displayed by the contractors in the development of the exceedingly mobile drilling rig, an interesting feature was the length of steel used in drilling. Owing to the break of the rock, which in some cases left a pronounced slope from top to bottom of the main bench, 10- and 12-foot starters in the top holes were often necessary. The crews became exceedingly skillful in the handling of these lengths of steel, and also in their other operations.

During December and January, when eight faces in the bench headings were being worked, ten Ingersoll-Rand drill-steel sharpeners were in service. Shops were operated continuously in three shifts, and 30 tool sharpeners and 30 helpers were employed. These shops also cared for miscellaneous sharpening, in connection with the scaling of the canyon walls, railroad-bench and open-cut excavation, and other outside work. It is interesting to note that 23-foot steel was used on the drilling jumbos on the bench and invert headings. This accounts, in a large measure, for the excellent footage made per round. It required 97 tons of steel to equip

**Above—Gantry crane used for pouring invert concrete.**

**Right—Forms ready for placing the concrete lining in invert.**

all the jumbos during this period. Each jumbo, on entering the tunnels to commence drilling, carried 25 sets of steel, which weighed 5½ tons. Each set consisted of from six to nine pieces, depending on the length of the starters.

The steel was changed on an average after each round—the sharpened steel being brought forward to the jumbo on 1-ton trucks when needed. Hollow, round, 1¼-inch drill steel was used throughout the job; and it was estimated that the loss due to wear, sharpening, breakage, and other causes was .32 pound of steel per cubic yard of rock removed.

The average time required to back the jumbo to the face, jack it into position, attach the air, water, and electric lighting connections, and point the drills, was twenty minutes. The average drilling time on the bench heading was four hours, and the average footage per round was 16 feet. The best progress on any bench heading was made between February 1 and 8, 1932, in the upper heading of Tunnel No. 4, when 280 linear feet was driven in sixteen rounds—an average of 17.5 feet per round. The best record made in any 24-hour period on all bench excavating was 256 linear feet in eight headings on January 20, 1932. The best individual record for one heading was 46 feet.

PAUL GUINN
*Assistant Tunnel Superintendent*

Mucking was carried on in the enlarging operations by Marion Type 490 one-hundred-ton electric shovels loading directly into dump trucks. These shovels came onto the job equipped with 2¼-cubic-yard rock dippers; but after some experimenting a 3½-cubic-yard dipper was installed by the contractors. Eight of these shovels were employed, one for each heading. Five drilling jumbos were in use, four of which alternated between two headings, with one in reserve. The work was so divided that one set of dump trucks served two headings in the lower portals, with approximately 25 trucks on each side of the river. At the upper portals, about 50 trucks of larger capacity were in use, serving all four portals as mucking operations required. The trucks were equipped with specially designed rock bodies, as can be seen in the accompanying illustrations. They were of various makes and types with capacities of from 7 to 14 cubic yards.

The muck was at first removed by hand from the small bench, which was carried one round ahead of the main bench. Because of the necessary length of the bench, this operation was tedious. A caterpillar 30 tractor, equipped with a bulldozer, was later installed in the top headings, and muck from the bench was pushed off and removed by the shovel from below. Mucking operations at the main bench were facilitated by the use of caterpillar tractors equipped with a bulldozer in front and a cowdozer in the rear. After a shot was fired, the bulldozer concentrated scattered muck at the base of the muck pile. Several times during the mucking operations the bulldozer was called into use to bunch the muck for ease of handling by the shovel. The cowdozer was called into play to scrape the muck away from the bench and to keep the tunnel floor alignment true. This also facilitated the alignment of the drilling jumbo at the face of the heading.

During the firing, the shovels were moved a few hundred feet back from the face. These electric shovels made an excellent performance record throughout the operations, with remarkably few shutdowns for repairs. An average of 110 cubic yards per hour was maintained, and individual performances were as high as 200 cubic yards per hour. The average mucking time, after each round, was nine hours, and approximately 1,000 cubic yards was broken with each round. During January, 1932, as much as 16,000 cubic yards, solid measurement, was removed daily from the tunnels and hauled to spoil dumps.

The problem of the disposal of the vast amount of broken rock, or muck, was one of the most difficult faced by Six Companies

B. A. PETERS LEIGH CAIRNS PETE HANSON FRANK BRYANT

**These four "master muck movers" served as traffic managers for blasted materials**

Incorporated. Under the specifications of the contract, no material could be dumped into the river. Disposal areas were designated in side canyons; and where these were not available the muck had to be hauled up the side of the canyon walls on roads cut from precipitous cliffs. Steep grades of necessity prevailed. The distance that muck was hauled at the lower portals was one-half to one mile. To reach one disposal area, a 380-foot truck tunnel was necessary. At the upper portals, muck was hauled in dump trucks for a part of the time and dumped directly into 30-cubic-yard side-dump railway cars and hauled upstream approximately two miles to be used in widening the railway grade into the canyon.

The rock encountered in driving the diversion tunnels has been characterized as ideal for these operations. It is volcanic in origin, and geologically determined andesite tuff breccia. The rock is easily drilled; and when properly loaded breaks so that it may be conveniently handled by shovels. The rock is referred to by mining men as "dead". It requires accurate placing of the holes, but breaks remarkably true. No major faults were encountered, and such small seams as developed were all closed. No heavy ground was met, and no spalling or air-slacking developed—the tunnels standing through their entire lengths without timber support of any description. No water was found in any of the top headings, and only a small amount during the excavation of the bench headings. Some seepages occurred during the excavation of the invert, especially near the portals, but these will be grouted off during the lining operations. Measurements taken after the completion of the trimming and scaling operations disclosed that, despite the rapidity with which the tunnels were driven, the average overbreak in the 56-foot section was only 7 inches.

Owing to special conditions that developed where open-cut excavations were necessitated and scaling operations were required above the tunnel portals, the dates of starting the enlarging operations of the tunnels varied from September 21, 1931, to December 27, 1931. The following tabulation showing the rate of progress is extremely interesting:

| | Footage Progress | Faces Worked | Cu. Yds. Excavated |
|---|---|---|---|
| September....... | 60.5 | 1 | 4,313 |
| October......... | 625 | 3 | 42,887 |
| November....... | 1,809 | 6 | 122,455 |
| December....... | 3,848 | 8 | 255,866 |
| January......... | 6,773 | 8 | 447,018 |
| February........ | 1,958 | 5 | 83,028 |
| March.......... | 1,330 | 2 | 87,780 |

On March 28, 1932, there remained to be driven 205 feet of bench heading out of a total of 15,909.28. The estimated total excavation from the four diversion tunnels is 1,451,369 cubic yards, there being 91.225 cubic yards per linear foot in an ideal section of the 56-foot bore, made up as follows: Top heading, 5.333 cubic yards; bench heading, 66.002 cubic yards; and invert, 19.890 cubic yards.

TOM REGAN
*Assistant Tunnel Superintendent*

The first tunnel holed through on the bench excavation was No. 3, the one closer to the river on the Arizona side. This tunnel—3,560 feet long—is the shortest of the four; and the final shot was fired on January 30, 1932. A few days later, on February 3, Tunnel No. 2 on the Nevada side, 3,879 feet long, was holed through. Work was suspended on bench headings in the lower portal of Tunnel No. 4 on February 1, 1932; and then was started the remaining section of the invert arch necessary to complete the excavation from a horseshoe section to a complete circle. Immediately after the bench heading was finished, invert excavation was taken in hand on both upper and lower portals of Tunnels Nos. 2 and 3. The bench excavation was continued on the upper heading of Tunnel No. 4, which was holed through on March 3, 1932.

The final enlargement of the tunnels is being completed in two operations. While the invert section is being excavated, the trimming and scaling of the tunnel walls to remove the projecting rock is carried on. The removal of the invert section is a similar operation to that on the bench heading. The same drilling carriages with the top part removed and with two folding wings built on either side are used for this purpose. Drills are mounted on the wings, which form a bar, curved on a 28-foot radius, the whole of the invert section thus being drilled in one operation. During February, 1,993 linear feet of invert was removed, and in March, 5,692 feet. Six faces were worked, and 152,931 cubic yards of muck was removed out of an estimated total of 316,604 cubic yards.

The operation of trimming or scaling rocks projecting within the clearances allowed is accomplished by a trimming jumbo, and is carried on coincidently with, and a short distance ahead of, the invert excavation. A horseshoe-shaped steel framework is erected in each tunnel. This framework has an outside diameter of 50 feet, and is mounted on wheels traveling on 90-pound rails laid true to line and grade. Platforms are erected at different elevations on the framework, and drills are mounted on bars at various points. The projections within the minimum allowable clearance are thus easily determined, and, when necessary, short holes are drilled and protruding points blasted on to

"SI" BOUSE, *Master Mechanic* "MORT" LEDERER, *Sup't Motive Equipment* C. A. HARRIS, *Chief Electrician*

**Three important personages in the tunnel-driving organization.**

the floor, where the muck is removed by the invert operation, which follows.

Ventilation was supplied in the adits and top headings by Roots blowers, discharging air through 18-inch pipes at the rate of 8,000 cubic feet per minute. During the driving of the main bench headings, a strong current of fresh air was drawn in through the 10x8-foot adits, forced through the 12x12-foot top headings, and discharged into the enlarged tunnels by pressure fans with a capacity of 35,000 to 120,000 cubic feet per minute—the air being introduced in the top of each tunnel, aiding the natural air currents from the portals.

Tests showed that there was a natural air movement of 1 to 2 feet per second from the portal toward the face in the bottom half of each large tunnel, and one of similar velocity in the opposite direction in the upper half. These natural currents were the result of a temperature difference between the atmosphere outside and the warm rock inside. A striking example was the relatively rapid up-draft always present at the face of a newly broken down muck pile.

The introduction of fresh air through the top headings accelerated the natural air currents and maintained a cool, clean, and pleasant working condition for the muckers. Tests showed that, under working conditions, activities could be resumed in five minutes after blasting with perfect safety and comfort; and the first truckload often went out fifteen minutes after the shot. Natural convection took the powder smoke to the vaulted roof and allowed men and equipment to move in under it. Then the accumulated gases were forced out by the combined action of natural and forced draft.

All the workings, involving truck operations, were thoroughly and periodically checked for air pollution. The degree of vitiation, particularly with relation to carbon-monoxide gas, was insignificant in comparison to that of vehicular tunnels, particularly where these have a definite sag, such as in the Holland tunnel.

During February, 1932, a flash flood in the Colorado River, amounting to approximately 50,000 cubic feet per second, topped the temporary embankments and flooded the tunnels. Damage was nominal; and work was suspended for several days while the tunnels were pumped out and the deposit of slime and silt was removed.

During the driving of the main bench headings, a crew of approximately 80 men was required in each heading. The crew on the drilling jumbo consisted of 22 miners, 21 chuck tenders, five nippers, one safety miner, and one drilling foreman or shifter. In addition, two crews of fifteen men each were engaged in drilling the wings on either side of the top heading. The mucking crew consisted of a shovel operator, and oiler, and a pitman. In addition to the foregoing, electricians, pumpmen, powermen, and superintendents operated in more than one heading. The daily wages of the jumbo and mucking crews was as follows: miners, $5.60; chuck tenders and nippers, $5; shovel operators, $10; oilers and pitmen, $5.

JACK LAMEY
*Assistant Tunnel Superintendent*

C. T. HARGROVES
*Assistant Tunnel Superintendent*

# I-R Blacksmith Shops at the Hoover Dam

*Above, one of the first blacksmith shops to be put in operation. Left, a close-up view of another shop, showing sharpener and oil furnace in operation. Below, the No. 50 sharpener that is so widely used in all parts of the world.*

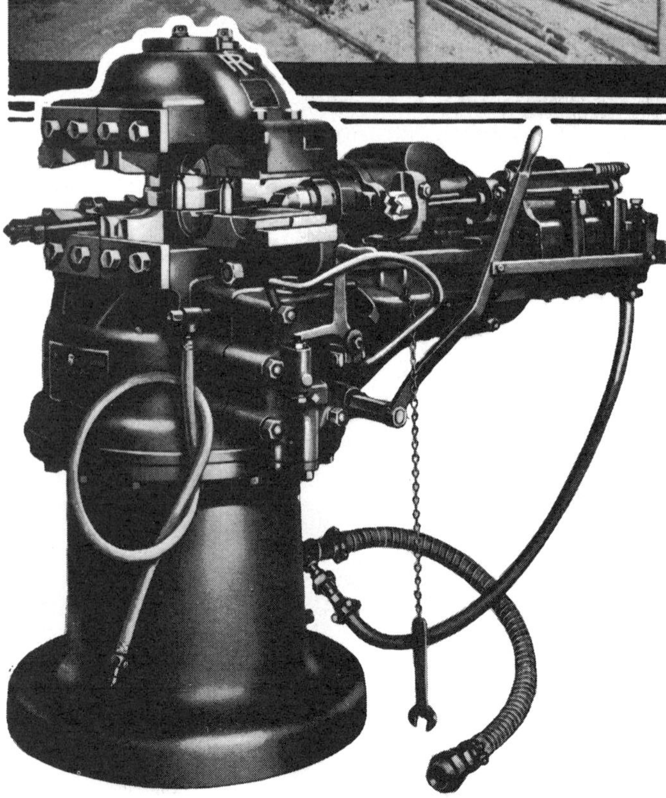

AT the peak of the drilling operations at the Hoover Dam, there were as many as ten completely outfitted blacksmith shops for reconditioning the great amount of drill steel required for the job. Many thousands of bits and shanks were reconditioned every 24 hours during January, 1932, when the work in the tunnels was being pushed at top speed.

I-R sharpening equipment was used in each of these blacksmith shops. It included a No. 50 drill-steel sharpener, a No. 26 oil furnace for heating the steel, a No. 8 grinder for squaring shanks, and a No. 6 drill for opening plugged steel.

**Ingersoll-Rand**

**Laying an air line across the Colorado River on a suspension foot bridge.**

**Erecting a Class PRE compressor—one of five such machines on the job.**

# Construction of the Hoover Dam

*Compressed Air Plays a Part of Vital Importance in this Huge Undertaking*

**COPELAND LAKE**

THE gigantic engineering and construction drama being enacted within the sheer walls of Black Canyon calls for many players, both human and mechanical. Various tools and machines of diverse sorts have already assumed important roles, and many others will attain prominence as the spectacle moves along toward its finale. No one of them, or no dozen of them for that matter, can, however, be given sole credit for the remarkable progress being made. All are important, not of themselves, but in conjunction with the others. Teamwork is paramount, but each contributor to the unison of effort that gets things done is as vital to the general scheme of operations as the strength of each individual link is to the power of a chain.

Keeping these things in mind, it is still possible to truthfully say that the rock drill, more than any other one class of tool or machine, has been the principal performer during the early stages of the work. It came into play at the very outset, and has been continually in the forefront of things ever since. Thousands of cubic yards of rock had to be removed by drilling and blasting in carving out the highways and railroad lines required to reach the canyon proper in order that work might begin in earnest. Long before these avenues of approach were in service, however, the advance guard of dam builders had floated down the river itself and established a meager foothold from which to start the first diversion tunnel adit, as related in the previous article. Soon afterwards the first of the scores of "scalers" with their "Jackhamers" took up precarious perches at various points along the canyon walls to clear away loose and projecting rocks and to open up tunnel portals and other excavations. Then came the actual drilling of the 56-foot diversion bores. Meanwhile, a great amount of miscellaneous drilling was going on, so that literally hundreds of drills were in use virtually all the time during the first year after Six Companies Incorporated took active hold. At one time as many as 1,200 men were engaged in operations having to do with drilling, blasting, and moving muck.

Since compressed air is the breath of life to the rock drill, it naturally follows that a compressor installation of considerable size was required. When the matter of such equipment was first taken under consideration it was estimated that approximately 25,000 cubic feet of air per minute would be needed, and plants aggregating that capacity were contemplated. Actually, however, the heaviest drilling schedule to be encountered during the life of the contract was carried out with a total air supply of 16,195 cubic feet per minute.

Because so much depended upon an adequate and sustained volume of compressed air during the initial stages of the work, Six Companies Incorporated weighed carefully the selection of the machines that were to be called upon to keep the program of rock drilling moving at a fast pace 24 hours a day with scarcely a break. In the course of the consideration it was decided that it would be advantageous to buy all the equipment required from one manufacturer so as to reduce negotiations to a minimum, to centralize the responsibility for installation and functioning of the compressors, to keep the necessary stock of spare parts low, and to simplify the duties of operation, care, and maintenance. It was also determined that it would be desirable to standardize, as far as possible, with respect to types or models and sizes. By so doing, spare parts would be largely interchangeable among the various units, and the problems of operation, care, and maintenance would become proportionately simpler.

Obviously, the primary demand was for compressors that would perform efficiently and that would, moreover, stand up under the rigorous service conditions that were bound to be imposed upon them. With a view to making a choice that would satisfy these vital demands, those charged with the responsibility of purchasing the air equipment gave close attention to the past records of the various makes of compressors which were being offered them.

In addition to the foregoing factors there was, however, another point which had to be taken into account in connection with the plans for the air plant. This had to do with the scheme which had been devised for cooling the concrete in the great dam which will some day stay the rush of the Colorado River. As is well known, the chemical processes which take place during the setting of concrete are accompanied by the generation of much heat. This heat has the effect of expanding the mass, which calls forth the reciprocal action of contraction when the heat is given up. When large blocks of concrete are poured at a time, this inevitable contraction produces fractures which weaken the structure. To obviate these difficulties, it is the universal practice to rear concrete dams on a sort of stagger system—that is, to build them progressively in sections or columns, which are carried upward by easy stages. Adjacent columns are poured at different intervals, and the intervening spaces are left open for sufficient periods to facilitate the cooling of the concrete by providing radiating

At the top is one of the diesel-engine-driven compressors used until electric power became available. Next below it and at the left are two Annis filters on the air intakes of compressors. At the right, and also below, are views in Compressor Plant No. 3. At the bottom are a No. 26 oil furnace and a No. 50 drill-steel sharpener in action.

surfaces exposed to the air on all four sides. In effect, the whole area to be concreted is checkerboarded, and the squares are built up a little at a time with no two squares having sides in common being poured simultaneously. Such a system insures a minimum of shrinkage and compensates for such contraction as does take place by setting up joints which can afterwards be sealed tight with grout.

As the gaps between alternating columns in the same horizontal plane cannot, under customary methods of procedure, be poured until the major portion of their contained heat has been dissipated, the progress on a dam as a whole is necessarily slow. On relatively small structures the sum total of these delays is not of much consequence, but in the case of the Hoover Dam, with its great monolith of 3,400,000 cubic yards of concrete, it would, in the aggregate, amount to a tremendous period of time (the estimate is 200 years and more) and have the result of prolonging the construction, correspondingly increasing the expense, and postponing the date when the structure would become of service and value.

Accordingly, in the interest of quickening natural processes, the Bureau of Reclamation engineers specified that artificial cooling of the concrete should be brought into play by the contractors. It is not our purpose to here describe how the dam will be assembled, but rather to sketch the essential features of the projected plan for cooling the concrete because of the bearing it had on the selection of the air-compressing equipment.

The specifications require that after any portion of the concrete in the dam (and tunnel plugs) has set for a minimum period of six days it shall be cooled by removing the excess heat above 72° F. This is to be done by circulating water of requisite coolness through the concrete by means of a system of pipes. This involves the provision and operation, by the contractor, of a complete refrigerating plant.

To carry out this scheme, there will be needed a network of 2-inch piping aggregating 800,000 linear feet, or roughly 150 miles. This will be assembled in the form of loops or series of loops ranging up to a maximum length of 1,600 feet. Where pipes cross contraction joints they will be linked by expansion-type couplings. This piping will be embedded in the concrete and will remain a permanent part of the dam structure.

When poured, the concrete will have a temperature somewhere between 40° and 100° F. Because of the wide seasonal range in the natural temperature at the dam site —from 25° to 125° F.—the amount of heat that will have to be extracted to lower the temperature of the concrete to the prescribed 72° will vary greatly as between summer and winter. The average temperature rise that will result from the setting of the concrete will be 40° above that at which it will be placed. The amount of heat that will have to be removed is about 700 B.T.U's per degree per cubic yard of concrete. The estimated quantity of heat to be extracted from each cubic yard of concrete during each month of the year, and the time that will be required to do this, are shown in a table, on a following page, which is from specifications prepared by the Bureau of Reclamation. It is to be noted that the actual amounts may vary considerably from the estimates in any one year, as the figures are based on average natural temperatures.

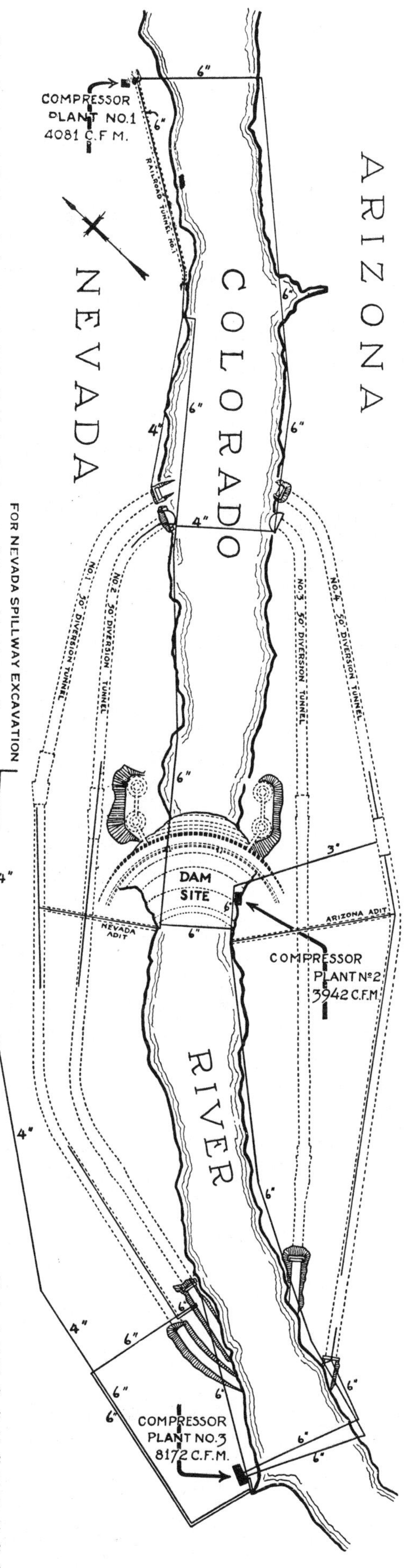

**The compressor plants and principal air delivery lines.**

It is provided that the refrigerating plant shall have a capacity sufficient to reduce from 47° to 40° the temperature of a flow of water of 2,100 gallons per minute. The water entering the cooling pipes shall not have a temperature lower than 35°, and may leave the embedded pipes at any temperature between 42° and 65°. The quantity of cooling water circulated may vary from 350 to 2,100 gallons per minute.

It can be realized from the foregoing facts that the contractors had in mind something besides the provision of an adequate air supply when they considered the purchase of compressors. They were concerned with obtaining machines which would not only prove efficient, reliable, and economical as air producers but which could also be readily converted, as needed, into refrigerating units.

After due investigation, they selected Ingersoll-Rand electric-driven compressors equipped with clearance-valve control. In all, eleven machines were purchased. Five of them are Class PRE-2 units, each of 26&16¼x18-inch size and having a piston displacement of 2,195 cubic feet per minute. The six others are the similar although smaller Type XRE-2 machines, each 20&12½x14 and with a piston displacement of 1,302 cubic feet per minute. All are driven by Westinghouse synchronous motors with belted exciters. When the time arrives to set up the cooling plant, the substitution of ammonia cylinders for the regular cylinders on a suitable number of these compressors will convert them into efficient refrigerating units. It is expected that only the XRE machines will be used for this purpose, and that not all of them will be required.

To supply air on the job, there were established three compressor plants, the locations of which are shown on the accompanying map, together with the principal air delivery lines. It will be noticed that although the plants were not close together—the extreme distance being some two miles between Plants Nos. 1 and 3—all were connected by the piping system.

As told in the previous article, Plant No. 1, at the upper end of the canyon, was the first installed. Until electric power became available in the early summer of 1931, the two Type XRE units which it then contained were operated as belted machines with power furnished by two Atlas diesel engines. Plant No. 2, near the adit leading into the diversion tunnels on the Arizona side of the river, was the second to be set up, and it consisted of two Class PRE machines. Plant No. 3, just below the lower portals of the Nevada diversion tunnels, was the last to be erected, but became the largest of the three, with a

Above is Black Canyon as viewed from the top of the Nevada cliff and downstream from the dam site. The benches at the river edges were built up of muck taken from the adits and the diversion tunnels. The dam will rise at about the line where the lowest bridge spans the Colorado.

Left—Compressor Plant No. 2 as it appeared from above during erection. In the upper left-hand corner can be seen both frames of one machine and one frame of another one. Below them and at the right are three portable compressors. This plant is at the right of the river in the picture above.

capacity of approximately 8,172 cubic feet per minute. Plant No. 2 had a capacity of 3,942 cubic feet per minute. Plant No. 1 had an initial capacity of 2,210 cubic feet per minute, which was later increased to 4,081 cubic feet per minute through the addition of a Class PRE machine.

With the exception of the two Type XRE machines in Plant No. 1, all the compressors were served by I-R Class HM tubular-type aftercoolers. These coolers added to the efficiency of the drills by eliminating troubles arising from moisture in the air lines. To further guard against preventable troubles in the air system, the compressor intakes were equipped with Annis air filters of the type which can be cleaned with an air blast without taking them out of service. Circulating water for the compressors was supplied by Cameron Motorpumps, which combine a motor and a centrifugal pump in one compact casing. Two such pumps were installed in each plant.

Since the several installations were of a temporary nature, and as the chief concern of the contractors was to get the compressors operating, no effort was made to erect model plants from the standpoint of appearance. The machines were set up in the open, and rough buildings were later constructed around them. The piping was chosen on the basis of sizes available rather than of which were preferable to make a neat-looking job. Six Companies Incorporated asked only that the machines function properly and keep the drills pounding away at the rock; and the record of the drilling accomplishments shows that they were given no cause for disappointment on this score. The air was discharged from the compressors at 105 pounds pressure, and was delivered to points of work on the spillways as much as a mile away at 80 to 85 pounds. Within the tunnels the pressure seldom was below 90 pounds.

*Snaking a portable up a steep slope.*

Most of the compressors were installed during the torrid season of 1931, at a time when it was impossible to handle metal in the sun with the bare hands. On more than one occasion gasoline, which was being used to clean small parts, caught fire through spontaneous combustion. The first two machines to be set up were floated down the river on barges. Rock had to be blasted away at the foot of a cliff to secure enough level space to erect them; and they were snaked up the steep slope from the river with block and tackle. Plant No. 2 occupied ground reclaimed from the river by filling in at the base of the cliff with spoils from the first of the adits, which was started with air from portable compressors. The units installed there also were floated down the river.

During the period of approximately a year that they furnished air for the tunnel drill

Heat Units to be Extracted from Concrete to Accelerate Setting

| | Mean Monthly Temperature | Maximum Temperature of Concrete | Heat to be Extracted | B.T.U.'s to be Extracted per Cu. Yd. | Cooling Water Must be Applied |
|---|---|---|---|---|---|
| | *Fahrenheit* | *Fahrenheit* | *Fahrenheit* | | *Months* |
| January | 52.0° | 92.0° | 20.3° | 14,200 | 1.14 |
| February | 57.2° | 97.2° | 25.5° | 17,900 | 1.33 |
| March | 63.6° | 103.6° | 31.9° | 22,400 | 1.60 |
| April | 71.2° | 111.2° | 39.5° | 27,700 | 1.83 |
| May | 78.6° | 118.6° | 46.9° | 32,900 | 2.06 |
| June | 87.6° | 127.6° | 55.9° | 39,200 | 2.28 |
| July | 93.8° | 133.8° | 62.1° | 43,500 | 2.40 |
| August | 91.9° | 131.9° | 60.2° | 42,200 | 2.37 |
| September | 83.0° | 123.0° | 51.3° | 35,900 | 2.17 |
| October | 70.8° | 110.8° | 39.1° | 27,400 | 1.83 |
| November | 59.4° | 99.4° | 27.7° | 19,400 | 1.39 |
| December | 51.5° | 91.5° | 19.8° | 13,900 | 1.10 |
| Average | 71.7° | 111.7° | 40.0° | 28,000 | 1.79 |

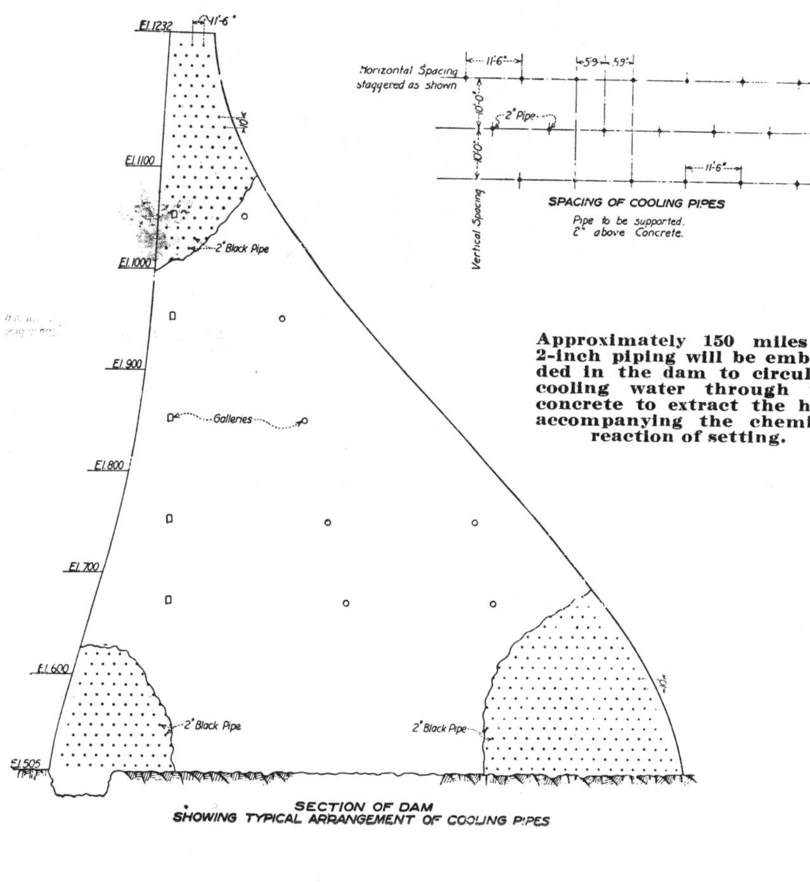


**Approximately 150 miles of 2-inch piping will be embedded in the dam to circulate cooling water through the concrete to extract the heat accompanying the chemical reaction of setting.**

ing, the compressors were called upon to operate through a temperature range from around 125° in July to as low as 14° during the winter. Throughout virtually all this time they ran 24 hours a day and seven days a week, and remained in service weeks at a stretch without shutting down. No repairs were required and the valves were cleaned only once. This is considered a remarkable record for compressors operating under such climatic conditions.

In the foregoing narrative, and in the last previous installment in this series of articles, we stressed the part that rock drills have played and will continue to play during what might be termed the preparatory stage of operations. While it is true that drilling constitutes the major use of compressed air, there are so many other ways in which this medium of power transmission is being employed that our present account should at least mention them if it is truly to portray the importance of compressed air in connection with this epoch-making enterprise. As future articles will deal with some of these functions more in detail, it is our present purpose merely to set them down for the sake of completing the record of the principal services of pneumatic power.

Seldom, if ever, has one contracting job afforded so many or such diverse applications of compressed air. An unofficial but probably quite accurate count reveals that there are approximately 500 individual tools or machines which are driven by air. The list is growing all the while, and many additional uses will no doubt be found before the contract is completed.

Besides the great number of "Jackhamers" and drifter-type drills which have been mentioned, there are about 25 stoper drills in service. These are Ingersoll-Rand CCW-11 and R-51 types, and are used principally in raising from the diversion tunnels on either side of the river the inclined shafts which will serve as spillways to take care of excess water in the dam without the necessity of running it over the top of the structure.

There are 25 Ingersoll-Rand Type CC-45 paving breakers and ten air-operated hoists at work. Both of these classes of machines perform a miscellany of services. Smaller penumatic tools abound, among them being twelve clay diggers and varying numbers of drills, riveting hammers, chipping hammers, grinders, etc. All are of Ingersoll-Rand manufacture.

**"Jackhamer" men clearing the area where the 56-foot spillway tunnel on the Nevada side will have its opening. This inclined waterway is being driven by raising from the diversion tunnel with which it will connect.**

Much air is, of course, consumed by the sharpeners, oil furnaces, shank and bit punches, and pedestal grinders required in connection with the reconditioning of the vast amount of drill steel and paving-breaker steel on the job. At the peak of the drilling operations, there were as many as ten completely outfitted blacksmith shops for the performance of these services. The equipment in each shop included a No. 50 drill-steel sharpener, a No. 26 oil furnace, a No. 8 pedestal grinder, and a No. 6 drill for drilling out plugged steels. Approximately 3,500 pieces of steel were reconditioned every 24 hours during January, 1932, when the work in the tunnels was being pushed at top speed. These blacksmith shops were moved about from time to time as the drilling progressed. Not all of them are in use now that the heavy drilling schedule has been completed.

Cameron pumps of several types and sizes are used at various points for the handling of water. Many of these are pneumatically operated. Air-driven Fuller-Kinyon pumps are employed to convey bulk cement from the box cars in which it is delivered to the top of the concrete mixing plant. Gates on the batchers, mixer hopper, and dump hopper in this plant are controlled by compressed air. Before placing the concrete lining in the invert of the diversion tunnels, it is necessary to scour the rock surfaces free of all dirt and pebbles. This is accomplished by means of compressed air used in conjunction with water.

All in all, there seems to be plenty of justification for the statement that compressed air is in the forefront of the indispensable agenices at Hoover Dam.

**Loading a gravel train at Gravel Plant. The Shay geared locomotive was formerly used on the Talmalpias Railway near San Francisco.**

**Dumping muck to widen the grade of the railroad in Black Canyon.**

**One of the five tunnels on the Government built railroad from Boulder City to the rim of Black Canyon.**

# Construction of the Hoover Dam

*An Account of the Extensive Railroad System and of the Important Work It is doing*

NORMAN S. GALLISON*

THE transportation of materials and equipment that must be moved in the course of building the Hoover Dam is so huge an undertaking as to constitute almost an industry in itself. It involves the construction by Six Companies Inc. of twenty miles of railroad at a cost of nearly $1,000,000, as well as the operation of this system and a 9½-mile line built by the Government—close to 30 miles in all. It is estimated that $2,000,000 will be spent for operation, making the total cost of railroading activities to the contractor $3,000,000, or about 6 per cent of the bid price on the entire work. This does not, of course, take into account the expenditure of $455,509 for the construction of the Government line.

Over these rails during the next five or six years there will move a volume of freight greater than that on any main-line railroad in the West. Carefully compiled figures indicate that 33,000,000 tons of live load will be carried, and that this will result in the movement of approximately 440,000,000 ton-miles of dead and live load combined. The locomotives will travel an aggregate distance of 700,000 miles and will haul 63,000 trains whose combined 600,000 cars would make a solid line 4,500 miles long.

Virtually all of this railroad is built and in operation, in fact, thousands of tons of materials have already been transported. Like all other phases of the contract, the railroad runs 24 hours a day, seven days a week. The operating force numbers 71 men, and 60 others are engaged in track construction.

The accompanying map shows the railroad lines; but it furnishes little evidence of the difficulties that were met in their construction—difficulties that arose from the rocky and precipitous character of the surface in the vicinity of the dam site and from the great difference in elevation of the points that had to be connected.

Before the award of the contract was made to Six Companies Inc., the Union Pacific Railroad had built a 23-mile line extending from a point near Las Vegas, on its Salt Lake-Los Angeles system, to the site where Boulder City was later to rise like a mushroom. A 400-car switching yard was provided, as were also sidings to the locations that had been selected for warehouses and other structures that would require direct trackage. The Government had also placed a contract for the construction of tracks from Boulder City to the Nevada rim of Black Canyon directly above the dam site. Upon its completion, this line was turned over to Six Companies Inc. for operation. Its terminus has now been designated Himix, as it is there that the contractors will locate the high-level concrete mixing plant which will provide concrete for all that portion of the dam and its appurtenances extending above elevation 720 feet.

At a point 6.25 miles out from Boulder City on this Government-built railroad, the dam contractors started work on the remainder of the system that would be required. This junction point was at first known as Government Junction, but has more recently been named Lawler, after H. J. Lawler, a director of the contracting firm. All of this

*Public and Press Relations Division, Six Companies Inc.

Above—View of the railroad yard at Gravel Plant. Nearly 500,000 tons of raw concrete aggregates are in the background. Below—Track-laying crew of the subcontractors at work on Six Companies Inc. railroad in Hemenway Wash.

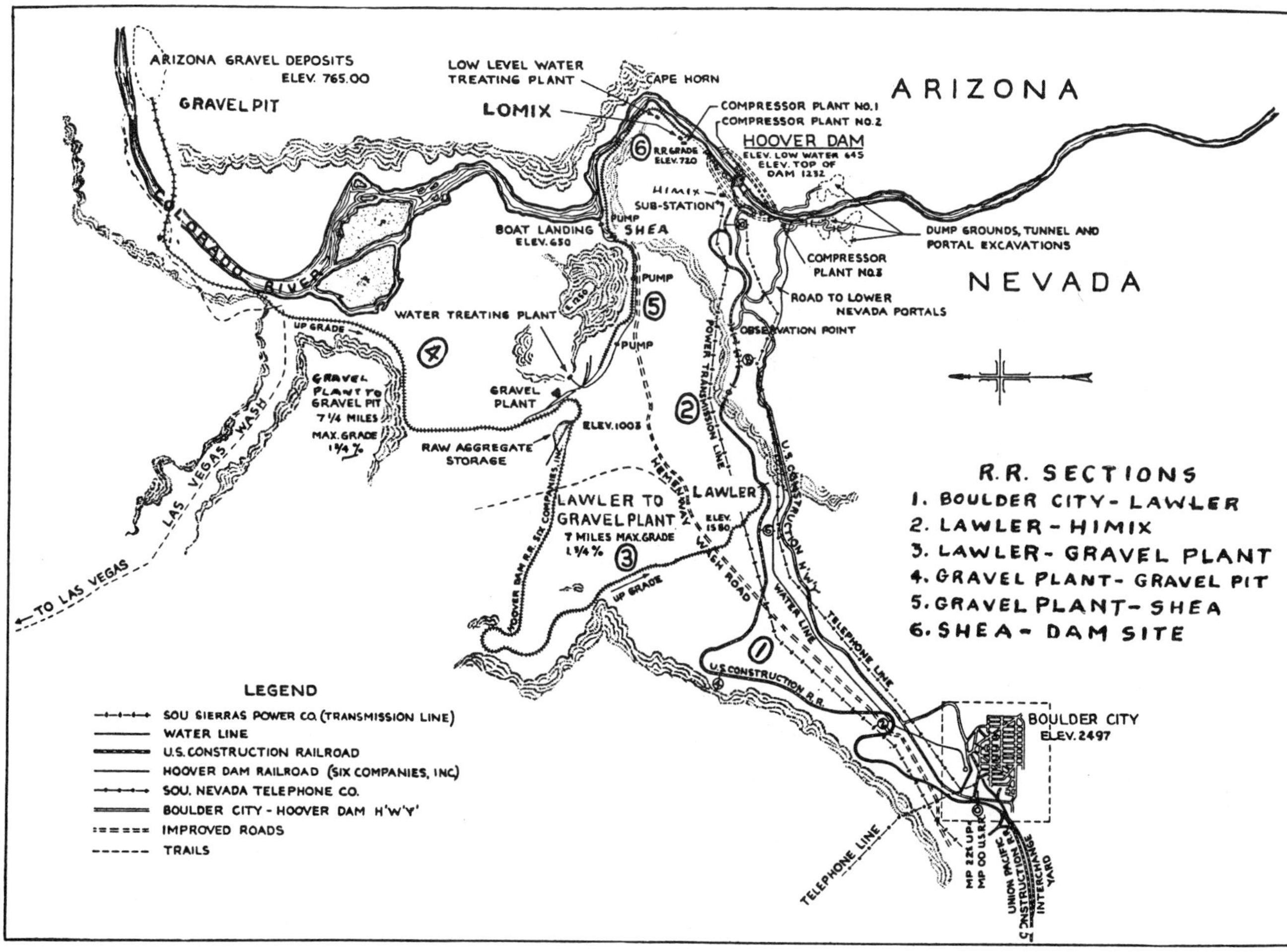


**Sketch map of the area affected by the Hoover Dam construction activities showing the railroad system, principal highways, power lines, and other structures provided to facilitate the rearing of the dam.**

railroad is of standard-gauge, 90-pound-rail construction, with Oregon fir ties laid sixteen to the rail length and ballasted. Except for that portion extending upstream from the dam site for three miles along the canyon wall, the line was constructed by subcontractors.

The section from Lawler to Gravel Plant is almost seven miles long, and descends in that distance from an elevation of 1,580 feet to one of 1,003 feet. Its maximum grade is 1.75 per cent. Gravel Plant is itself a junction, and is both figuratively and literally the center of the railroading operations. From this station, which is situated in Hemenway Wash about two miles from the river, the line runs upstream seven miles to Gravel Pit station at the Arizona Gravel Deposits and downstream five miles to the dam site. The upstream section crosses the Colorado about a mile and a half from the gravel deposits, this being the only portion of the trackage that is in Arizona. The rails descend from Gravel Plant to the river bank on a maximum grade of 1.75 per cent, reaching approximate water level in 3½ miles and then following the stream to their objective. The terminus at the gravel pits is at elevation 765.

The river is crossed on a wood-pile-and-trestle bridge 1,140 feet long. All concrete aggregates to be used in the construction activities at the dam will move over this span, and it is estimated this will involve 60,000,000 live ton-miles of operation. A suspension bridge supporting a belt conveyor was originally planned for carrying these aggregates across the stream, it being feared that a railroad trestle would be washed out by the first flood. Such a conveyor system would have involved additional handling of the gravel, however, and a comparison of cost estimates showed that the saving possible by loading cars directly at the pit would be sufficient to replace the trestle as many as four times during the life of the contract, should that become necessary. A railroad bridge was accordingly built; and, to minimize the loss in case of flood, it was anchored at each end with a large cable.

**Railroad equipment outside shops.**

From Gravel Plant toward the dam site, the railroad follows Hemenway Wash on a 3.4 per cent maximum grade a distance of two miles to Shea, a station named for Charles A. Shea, director of construction for the entire Hoover Dam job. This station is at the upper end of Black Canyon; and from that point downstream three miles to the dam site the line is known as the Canyon Railroad. This section was built by Six Companies Inc., and was the most costly of all the railroad construction because the rails skirt the canyon wall throughout their course. The excavation was almost entirely in rock; and two tunnels, each more than 1,000 feet long, had to be driven. From Shea to the terminus, the tracks run level at elevation 720. The lower end, above the dam site, is 80 feet higher than the river. For two miles between Shea and Lomix, where the low-level concrete mixing plant is located, the line is double tracked. This railroad system, which ranges in elevation from 2,497 at Boulder City to 720 in the canyon, comprises 36.7 miles of trackage, including double track, yard track, and sidings. A shop building 252 feet long has been constructed in Hemenway Wash and equipped for the care and maintenance of rolling stock.

**Cutting a shelf for rails in the perpendicular wall of Black Canyon. A 1,000-foot tunnel was bored through the rock at the right. The fill at the bottom was made with muck from the adit to the diversion tunnels on the Nevada side.**

For operating purposes, the line is divided into three sections. The first subdivision extends from Boulder City to Lomix, the second from Gravel Plant to Gravel Pit, and the third from Lawler to Himix. The system is run under the standard railroad code, the Union Pacific book of rules being followed in general owing to the fact that the crews work in and out of the interchange yard with that line at Boulder City. A single train-order form is being used. Dispatching is done by telephone.

Thirteen locomotives are in use. Three Mikado type and four consolidated-type steam units were purchased from the Union Pacific. There are also in service a 10-wheel, 70-ton steam locomotive, a Shay 40-ton steam locomotive, and four Plymouth 30-ton gasoline locomotives. Rolling stock consists of 50 bottom-dump hopper cars, 34 new-style 30-cubic-yard Western dump cars, 32 old-style 30-cubic-yard Western air-dump cars, six flat cars, and one tank car. Miscellaneous equipment includes one American gasoline locomotive crane, two Diesel Industrial Brownhoists, and one Jordan spreader.

The principal transportation problems involved in the Hoover Dam contract are: First, moving excavated materials from tunnels, dam-foundation excavations, and coffer-dam excavations to disposal areas; second, moving cement, sand, and gravel for concrete; and, third, moving miscellaneous construction materials and Government materials consigned to the power house. This class includes mechanical and electrical machinery.

Much of the materials in the first class has already been moved. Of the total amount of muck moving involved in the consummation of the contract, approximately three-fourths will be handled by trucks and one-fourth by railroad. All the second and third classes of materials will be transported by rail. Even though most of the material excavated from the diversion tunnels and related works was truck hauled, nevertheless vast quantities of it were carried by railroad. In general, excavated materials from the lower end of the working zone are handled in trucks and those from the upper end are transported by rail an average distance of 2.32 miles to disposal areas in Hemenway Warh. It is estimated that, during the life of the contract, a total of 2,933,743 tons of muck will be moved by rail. As many as 600 cars, containing 12,000 cubic yards of dirt and rock, have been hauled over the Canyon Railroad in a day. The schedule drawn up for the year 1932 calls for the haulage by rail of an average of 2,300 cubic yards of muck daily.

While the items included under the first and third headings are important from the railroad standpoint, the transporting of the materials for concrete will overshadow them. Sufficient sand, gravel, and cement must be hauled to make 4,293,400 cubic yards of concrete, all of which will be poured during the six years from 1932 to 1937, inclusive. During the two years 1935 and 1936, more than 3,250,000 cubic yards will be prepared and used. The schedule fixes July, 1935, as the probable maximum month, and estimates that 171,000 cubic yards will be poured during its 31-day span. Concreting of the diversion tunnels and related structures began last spring under a schedule which calls for the pouring of from 2,000 to 33,000 cubic yards monthly during the remainder of the year.

The sand and gravel which are requisite ingredients in the concrete will all come from the Arizona Gravel Deposits; and it will be necessary to transport 8,586,000 tons of these materials an average distance of 7.25 miles to the gravel plant, where they will be screen-

**Gravel Plant station, the "union depot" of the line.**

**The blast which put an end to "Cape Horn", a jutting cliff that impeded railroad construction at the upper end of Black Canyon. Eight tons of dynamite, loaded in scores of drill holes, brought down 160,000 cubic yards of rock.**

ed and washed. Moreover, it will be necessary to have all the needed aggregates out of the pits by 1936, because by that time the gravel deposits will be under the water that will have been backed up by the portion of the dam then in place. It is estimated that 5,618,000 tons or, roughly, three-fourths of the total supply will be moved by December, 1934. These materials are being transported to the gravel washing plant and run through it as fast as it will handle them. They are then being placed in stock piles to be used as required. The gravel treating plant and the stock-pile areas are high enough so that they will still be above the shore of the lake that will have been created by 1936.

**A muck train at Lomix, a mile above the dam site.**

The moving of the sand and gravel from the stock piles to the two concrete mixing plants will be another major railroading task. The quantity to be hauled to the low-level mixing plant, a distance of 4.75 miles, is 3,033,730 tons; and that to the high-level plant, a distance of 10.4 miles, is 5,124,730 tons. In addition, all the cement required will have to be transported from Boulder City. This will involve moving 374,586 tons a distance of eighteen miles to Lomix and 632,646 tons a distance of 9.65 miles to Himix. This cement will arrive in bulk in railway cars, which will be emptied at the mixing plants and then returned to Boulder City. At the height of concreting operations, the Himix plant will require 25 carloads or 1,150 tons a day. This is equivalent to 6,250 barrels. The Lomix plant will need about one-third this quantity at its peak activity. Miscellaneous materials such as reinforcing steel, structural steel, high-pressure gates, valves, pipes, fittings, and machinery for the power house will have to be moved from Boulder City down to the dam site and will aggregate 400,000 tons.

Gravel and sand hauling is being done by trains consisting of ten cars. A 5-cubic-yard Marion dragline loads the materials at the pit. Although the cars are rated at 30 cubic yards capacity, they are carrying an average of 35 cubic yards. The material weighs about 3,800 pounds to the cubic yard, so that 700 tons are transported on each train trip. En route, the locomotive must negotiate grades up to 1.75 per cent. A round trip requires about 2½ hours. During the past summer from 200 to 250 cars a day were being delivered to the gravel plant. Some of the screened and washed aggregates are now being used for the concrete lining of the diversion tunnels, which calls for their transportation from Gravel Plant to Lomix.

From the foregoing account it can be gathered that Six Companies Inc. have tackled their railroading problem with the same foresight, thoroughness, and engineering soundness that have characterized their other manifold activities pertaining to the record-breaking Hoover Dam undertaking. As yet the time of really heavy traffic has not arrived; but the facilities are ready to handle it when it does come. Meanwhile, crews are limbering up for the days when they will haul a volume of freight that might arouse the envy of any railroad executive in the country.

The Six Companies Inc. railroad is being operated under the supervision of T. M. Price, superintendent. Mr. Price designed and directed the construction of the gravel-screening plant. For a number of years he has been with the Kaiser Paving Company, one of the member firms of the Hoover Dam coalition. G. A. Allen is trainmaster and chief dispatcher. L. A. Grubbs is dispatcher on what is known in railroad parlance as the second trick, and T. J. Kelly serves in the same capacity on the third trick.

# Construction of the Hoover Dam

*A Description of the Methods of Obtaining and Preparing the Aggregates for the 4,400,000 Cubic Yards of Concrete to be Poured*

ALLEN S. PARK

THE sand, gravel, and stone for the 4,400,000 cubic yards of concrete that will be used to construct the Hoover Dam and its appurtenant works will all come from the Arizona Gravel Deposits, a 30-foot bed of alluvial material located about six miles in an air line upstream from the dam site. This lense-like accumulation of silt, sand, gravel, and bowlders was laid down in past years by the Colorado River. Thus the tempestuous stream is in a sense a copartner in the herculean task of bringing it under control. Not only did it oblige man by cutting a deep gash through the lava flows that provides a rock-ribbed anchorage for the great monolith of concrete that will be inserted there, but, as though to make an additional magnanimous gesture, it carried down in countless flood times and left near at hand much of the material that will go to fabricate that concrete plug. In this manner, perhaps, Nature requites for her ravages.

When the Government elected to build the dam, one of the multiplicity of duties that fell to the lot of the Bureau of Reclamation Engineers was that of finding a source of suitable concrete aggregates. That search was carried on actively for a number of months, and many possible sites were examined and reported upon. The Arizona deposits were found to be the nearest source of supply that would yield the character of materials wanted. Vast quantities of rock—hard, unweathered, igneous rock—are being excavated as a preliminary to the actual building of the dam. Crushed, it would make what would normally be considered a good material for concrete; but in the interest of providing the strongest and safest structure obtainable, the Government specified that river gravel deposits should be used for the purpose. Fortunately, these are not only close by but are present in the needed quantity. It is estimated that, after 24 inches of undesirable top material is stripped off, the 100-acre gravel bed will just about produce the 4,500,000 cubic yards of aggregates which will be required. The deposit is owned by the Government and was turned over to the contractors for use.

In their natural state, the aggregates are not pure enough to satisfy the rigid specifications which govern the making and placing of the concrete. It is required that the sand shall consist of "hard, dense, durable, uncoated, non-organic fragments", and that "it must be free from injurious amounts of dust, lumps, soft or flaky particles, shale, alkali, organic matter, loam, mica, or other deleterious substances." The clauses pertaining to the coarser aggregates are equally exacting. In the further interest of securing concrete which will stand through the years without decay or weakness, the specifications provide for grading the raw materials into sand, three sizes of gravel and one of cobbles, and for their segregating so that they may be used in precisely the proportions desired for making concrete for each of the several different purposes. It is also stipulated that the aggregates shall be washed prior to use—this treatment being designed to free them of adhering silt, sediment, organic matter, or other undesirable substances. The concrete, it may be stated here, must have a compressive strength of 2,500 pounds to the square inch where it is deposited in mass quantities and of 3,500 pounds per square inch for thin sections.

Pressing the button that started the plant on January 9, 1931. Left to Right—H. O. Watts, Southern Sierra Power Company; S. O. Harper, assistant chief engineer Bureau of Reclamation; operator of remote-control station; C. A. Shea, director in charge of construction Six Companies Inc.; F. T. Crowe, general superintendent; T. M. Price, assistant general superintendent in charge of aggregate production.

To prepare the aggregates in the manner designated and in the huge quantities required, Six Companies Inc. have expended some $450,000 in providing what is undoubtedly the largest screening and washing plant ever erected for such a purpose. The framework contains 350 tons of structural steel. This plant is situated in Hemenway Wash on flat ground about two miles from the river. Adjacent to it is Gravel Plant station on the contractors' railroad, a 3-way junction from which one rail line extends seven miles to the Arizona Gravel Deposits, another 4.7 miles to the dam site, and a third seven miles to Lawler, where it meets the Government built section from Boulder City to the Nevada Canyon rim above the dam site.

As set forth in the previous article, the gravel pit will be submerged by 1936, and it is accordingly necessary that all the aggregates required to complete the work be removed prior to that time. This circumstance and the fact that small quantities of concrete were needed during the early stages of the work led the contractors to provide facilities for treating the aggregates without delay. The screening and washing plant was started in November, 1931, and was operated for the first time on January 9 of this year. The first concrete was poured on March 5, 1932, to form a foundation for the trash rack at the inlet portal of one of the two diversion tunnels on the Nevada side.

The specifications permit the use of aggregate materials ranging up to 9 inches in section, and it is the function of the gravel plant to crush rocks above that size and to cleanse and segregate into the five prescribed sizes the materials under that maximum limit. As at present arranged, the plant performs these operations upon more than 600 tons of raw aggregates an hour, and it has been laid out so that its capacity can be increased to 1,000 tons an hour. Its essential units are a scalping station and crusher, four classification towers, a sand washer, a sand conveyor

Left—Loading raw aggregates at the Arizona Gravel Deposits.

Dumping a 30-cubic-yard car of aggregates at the screening plant.

Left—Live aggregate storage piles.

Below—General view of the gravel screening plant that will treat the aggregates for nearly 4,500,000 cubic yards of concrete.

Completed installation for storing sand, showing the concrete base and drain tile insets.

Below—The crushing plant.

The water clarifier during construction.

with automatic tripper, four live storage piles for gravel and cobbles on one side of the towers, and sand storage piles on the opposite side. The materials travel through the plant on a system of interrelated belt conveyors; and, after treatment, are lodged in their respective storage piles to await use. The entire system of handling, from pit to concrete mixing plants, operates with little manual supervision. In fact, fewer than 100 men are engaged in all its departments, including the transportation of the aggregates by railroad.

At the gravel deposits, a Marion Type 125 electrically operated dragline, equipped with a 5-cubic-yard bucket, loads material into Western air-dump cars which are rated at 30 cubic yards capacity but which commonly carry 35. Ten cars make up a train, which is drawn by steam locomotive to the gravel plant. A Bucyrus-43B dragline strips the unusable top portion from the deposit and casts it into adjacent areas from which the gravel has already been excavated. Operations at the pit are carried on in three shifts of eight hours each; and at the time this was written approximately 250 carloads, or 12,250 tons, were being hauled daily.

Upon reaching the plant, the cars are dumped either in raw storage piles or directly into depressed bunkers which have a combined capacity of 30 carloads of material. Under the bunkers are vibrating feeders which discharge the raw aggregates on to a 42-inch conveyor belt which runs in a concrete tunnel underneath. The belt transports them to the scalping station, where they are discharged into a cylindrical revolving screen 5 feet in diameter. The scalping screen passes material less than 9 inches through its perforations. Bowlders above that size are dumped on to a conveyor which carries them to an Allis-Chalmers gyratory crusher. Another conveyor returns the product of the crusher to the feed belt running to the scalping station.

The material which passes through the scalping screen is conveyed by a 36-inch belt to the first of the four classification towers, each 60 feet high. There the first of two vibrating screens allows aggregates less than 3 inches in size to pass through. The rejects, made up of cobbles from 3 to 9 inches in size, fall on to a transverse conveyor which transports them to a stock pile. The second screen takes out all sand less than ¼ inch in size

Below—The scalping tower.

Live storage sand pile a few weeks after the plant began operating.

Rake-type classifier producing sand

and passes it to chutes leading to separate treatment machines. The gravel which will not fall through this screen, and which consists of material ranging in size from 1/4 inch to 3 inches, is carried by a lateral conveyor to the second classification tower. The remainder of the gravel-treating process is a repetition of the procedure at the first tower. In the second tower, material coarser than 1 1/2 inches is removed and conveyed to stock piles. The third tower takes out the 3/4- to 1 1/2-inch sizes and passes them to stock piles. The final screening takes place in the fourth tower, from which the resulting 1/4-inch to 3/4-inch material is transported to stock piles. All screens except the scalping screen are installed in duplicate.

All stock piles are built up above concrete tunnels 11 feet high and 9 feet wide inside. Within these chambers are 30-inch conveyor belts leading to screens in the lower portions of the respective classification towers. Material being loaded out of a stock pile passes on to the belt through gates installed in the tunnel roof and controlled from the inside. It is then run over the screen to make certain that it contains no gravel smaller in size than that designated for that particular pile. This reclassification eliminates dust and debris that have blown into the gravel, as well as fragments that have broken from larger pieces during handling. The gravel that passes over the screen flows to a hopper and a 48-inch perforated conveyor, which loads it into bottom-dump railroad cars for transportation to the concrete mixing plant. While the gravel is in a stock pile it is wetted by a sprinkler system, and while it is moving on the conveyor it is played upon by water jets.

The sand which is laundered has water added to it at the first classification tower and enters a series of Dorr washers and classifiers. The first of these has an inclined bottom, and the sand is moved in stages from its lower to its upper end by a series of reciprocating rakes. The process is repeated in a second washer of the same type. As it reaches the plant, the sand usually contains an excess of material of 28 to 48 mesh size. This is segregated in the second washer, where from 1/3 to 1/2 of it is discarded. The remainder is recombined with the rest of the sand and passed through a bowl-type classifier for the removal of silt that escaped the previous treat-

**Top—The gravel plant, with screened gravel and sand in the foreground. Center—Geared-type Shay locomotive switching cars at the gravel plant. Bottom—How the gravel screening plant looks from the crusher bin. The screening towers are at the right.**

Plymouth locomotive switching a car at the Arizona Gravel Deposits during the early stage of operations.

ment. It is then dewatered in another rake-type machine, after which it is transported by belt conveyor through a tunnel to the opposite side of the railroad tracks. There it is delivered to an automatic tripper fitted with two stock-pile conveyors, one running in either direction and both perpendicular to the railroad tracks. From these stock piles the sand is loaded as needed into railroad cars by a railroad-type crane equipped with a clam-shell bucket. Tracks have been provided to permit loading from either side of the piles.

At its present capacity, the plant will supply material for the loading of a 50-ton car every four minutes. Storage space is available for 1,700 tons of cobbles, 1,500 tons each of the three sizes of gravel, and 22,000 tons of sand. By making comparatively slight alterations, speeding up the conveyor belts, and extending the tunnels and conveyors beneath the stock piles, the output can be increased to 1,000 tons an hour.

The entire plant is electrified, and more than 50 induction-type motors are installed. The main supply conveyor and the one serving the first classification tower are driven by 60-hp. motors, and the others by either 10- or 20-hp. units. These belts are all operated at constant speed through speed reducers from the motors. Most of the screens are operated by 5-hp. motors.

Water is an important agency, and it has to be pumped two miles from the Colorado River and raised approximately 415 feet. It is delivered through a 12-inch steel pipe line in three stages of pumping to an 800,000-gallon circular concrete reservoir, 115 feet in diameter and 15 feet high. This is located 130 feet higher than the gravel plant. The specifications require that the water used for washing the gravel be as low in turbidity as that which will go into the making of the concrete, the allowable maximum being 500 parts silt per 1,000,000. Raw river water contains approximately 36,000 parts silt per 1,000,000, but the suspended matter is of such a character that 98 per cent of it will settle out if the water is permitted to stand as much as three hours. This natural process of clarification is aided in the sedimentation tank by a Dorr traction thickener. The sludge is removed by diaphragm pumps. In connection with the sand washers, previously mentioned, there is a smaller sedimentation tank which receives the silt-laden water from these machines. It is equipped with a clarifier and a sludge pump. After treatment there the water flows to a sump and is then pumped to the large sedimentation tank which is situated on a hill overlooking the plant. The water system is a closed one, and the same supply is used over and over with only small additions from the river from time to time. In fact, of the 3,000 to 4,000 gallons of fresh water used in the plant per minute, approximately 20 per cent is pumped from the Colorado River and the remaining 80 per cent is water returned from the gravel plant.

The entire plant is controlled from a central switching station in the scalping tower. By means of 34 sets of push buttons, each motor and each step in the diversified operations can be individually controlled; and by throwing one master switch the entire plant can be shut down. From this same central station the nine gates or feeders in the bottoms of the supply bunkers are regulated by rheostat control, thereby governing the supply of aggregates reaching the primary conveyor and, incidentally, the output of the plant.

In its raw state, the gravel contains more or less organic matter and considerable silt. Before it enters the plant it has little of the appearance of good concrete-making material. After it has been treated, however, it looks altogether different. The several sizes all show up as clean, hard-rock fragments; and one has the feeling that concrete made from them will be as enduring as any structure ever assembled with the aid of human hands. Incidentally, the mineralogist might identify in these aggregates stones from various areas through which the Colorado River and its tributaries pour. It goes without saying that each of the states through which the river flows or which it touches has contributed to that great bar of gravel which was born of the river and which is soon to play an important part in harnessing this temperamental stream and thus turning it into a benefactor of mankind.

The structural-steel frame of the gravel plant was erected by the Pacific Iron & Steel Company of Los Angeles, Calif., and upwards of a dozen firms supplied essential equipment for it. The plant is being operated under the supervision of T. M. Price, who had charge of its design and construction. William Fudge and O. Haugen are his assistants. Government supervision of the production of aggregates is in the hands of O. G. Patch, who reports to Walker R. Young, construction engineer for the Boulder Canyon project.

# Construction of the Hoover Dam

The concrete mixing plant, with cars of bulk cement being unloaded by means of compressed air pumps.

*The Concrete Mixing Plant Surpasses in Capacity and Refinements any Previous Structure of Its Kind*

**C. H. VIVIAN**

THE BINDING together of the walls of Black Canyon with a great monolithic plug which will halt the rampages of the Colorado River and turn its moisture and energy to the benefits of mankind will call for pouring concrete in huge quantities and at a faster rate than has seldom before been attempted. The tentative schedule of construction drawn up by the Bureau of Reclamation prior to calling for bids allotted 32 months to the monumental task of placing the 3,400,000 cubic yards of mass concrete which will go into the dam proper. In view of what Six Companies Inc. have accomplished by way of speeding up other phases of the Hoover Dam work, it is possible that this estimated time will be shortened somewhat, but even the figure cited visualizes the formation and disposal of more than 3,500 cubic yards of concrete a day. This is equivalent to laying, in the course of 24 hours, a concrete roadway 10 inches thick, 20 feet wide, and more than a mile long, and keeping up this performance every day for 960 days. Even then, this comparison takes into account only the concrete that will enter into the dam structure itself and ignores the 1,000,000 cubic yards of additional concrete which will be used for lining the diversion tunnels and spillways and for constructing the power houses and other appurtenant works. Some of this will be handled at the same time that the dam is being poured. All told, as we have pointed out previously in this series of articles, Hoover Dam will entail the use of more concrete than entered into all the other dams constructed by the Bureau of Reclamation during its 29 years of existence prior to 1931.

Not only are the elements of time and quantity unparalleled, but, in addition, the specifications which govern the making and placing of the concrete are the most rigid and exacting ever to be applied to a large-scale construction undertaking. To further complicate the problem, the concrete must be poured under widely varying climatic conditions, as the inter-seasonal temperature range is commonly well over 100 degrees.

The general specifications of the Bureau of Reclamation provide that the proportions of the mix shall be such as to produce concrete which will have an ultimate compressive strength of not less than 2,500 pounds per square inch for the mass concrete of the dam and not less than 3,500 pounds per square inch for slabs, beams, and other thin reinforced members. In addition to its own painstaking and extensive researches, the Government commissioned experts outside of the Bureau of Reclamation to make thorough studies of such matters as the correct proportions for concrete ingredients, the effect of moisture changes on the shrinkage of concrete, the effect of heat developed in setting and methods of controlling it, as well as of the comparative merits of different types of cement. All of these things, and more, were done to insure that the best possible quality of concrete obtainable will be employed in this enormous artificial rock plug which will equalize the flow of America's most untamable river—a stream that has had a habit of changing within the span of a few days from a placid trickle to a raging, roily, destructive torrent of nearly a quarter of a million cubic feet per second.

To live up to the exacting regulations governing the preparation of the concrete, it devolved upon Six Companies Inc. to provide not only the largest mixing plant on record but also one which would exceed all previous ones as regards facilities for controlling the uniformity of its product and for insuring

**At the right is the mixing plant, with cars of aggregates over the track hoppers. At the bottom is the site where the plant now stands. The sketch shows the arrangement of the working parts.**

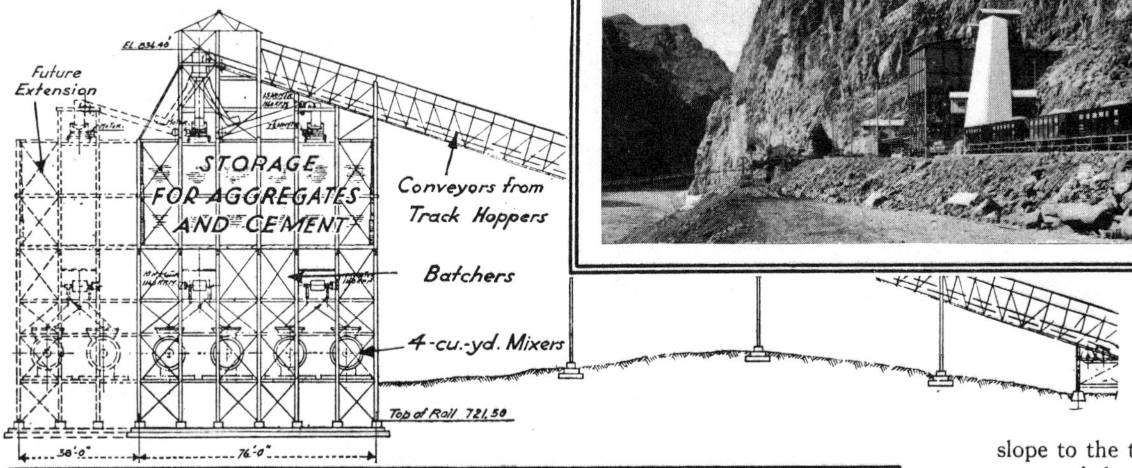

continuity of operation. Such a plant has now been built. It has been operating for several months past to supply the relatively small quantities of concrete being used for lining the diversion tunnels and for miscellaneous purposes. A few months hence, when the river bed has been dried up and excavations for the dam footings have been carried to bedrock, this plant will assume its heavier burden and become the center of activities.

Because of the steepness of the cliffs, the nearest location to the dam site that could be found for the mixing plant was 4,000 feet upstream, on the Nevada side of the river. At this point, designated Lomix on the contractors' canyon railroad system, it was possible, with the aid of rock drills and dynamite, sufficiently to enlarge the natural shelf to accommodate the structures required. From this plant at elevation 720 will come all the concrete for lining the diversion tunnels, for building the power-house foundations and for sundry other purposes, and two-thirds of the concrete for the dam. It is expected that the dam will be built up to elevation 930 or 935 from this set-up, or more than 200 feet higher than the high-water level of the river. After the work has reached that stage of completion, which will likely be sometime in 1935, the concrete plant will be dismantled, transported in units to Himix, on the rim of the Nevada cliff overlooking the dam site, and there reassembled. At that higher level the concrete for the remainder of the dam will be produced.

The mixing plant proper has ground dimensions of 78 by 118 feet and an over-all height of 117 feet, of which the upper 29 feet is a tipple for receiving and handling the aggregates. Upstream from this structure is a track hopper with compartments for the sand, three sizes of gravel, and cobbles, which are delivered there in 30-cubic-yard side-dump railroad cars from the gravel screening plant 4.7 miles distant by rail. To transport these aggregates to the mixing plant there are two inclined belt conveyors which rise on a 16° slope to the top of the tipple. The materials are carried an average distance of 450 feet by the belts, which have a combined capacity of 1,500 tons an hour.

In its present state, the mixing plant contains four 4-cubic-yard Smith tilting mixers, each driven by texropes from a 75-hp. motor. Ultimate plans call for an extension of the building to provide space for two additional mixers of the same kind and size. When this is done, the plant will be able to turn out a cubic yard of concrete every 9 seconds, or nearly 7 cubic yards a minute. This figure is based on each mixer producing 4 cubic yards of concrete every 3½ minutes. Working at this capacity the plant will be able to meet the estimated peak monthly requirement of 170,000 cubic yards by working 16 hours a day, 26 days a month.

The building which houses the mixing equipment is of steel construction and was fabricated and erected by McClintic-Marshall Corporation. Eight hundred tons of steel entered into it. It has laminated timber sides and partitions. Foundations for the mixing plant and the adjacent track hoppers involved the use of 3,000 cubic yards of concrete—which, incidentally, amounts to only about seven hours' output of the plant when it is operating at capacity. The structures were started in November, 1931, and the plant was ready for its test run near the close of February, 1932. The cost of the plant was approximately $250,000 and of the equipment $103,000.

Two railroad tracks run above the track hoppers for aggregates and provide room for spotting 10 cars at a time. Each hopper has a capacity of 300 cubic yards, or 10 carloads, of material. For loading aggregates onto the conveyor belts, there are 4 motor-operated chute gates of special design for each compartment, or 20 in all. Each belt can carry more than one class of material at a time, but in practice this is not done. At the top of the mixing plant, where a full view of all operations may be had, is a station where all the

**View downstream from the Arizona canyon rim showing the mixing plant and, left of it, the upstream portals of the two Nevada diversion tunnels.**

belts and gates may be controlled by push button. The system is connected electrically in such a manner that the stopping of any belt automatically stops all others behind it and closes the hopper gates. This eliminates the possibility of material piling up at any point because of the failure of any portion of the system to function.

In the upper part of the mixing plant are storage bins for each of the five classes of aggregates and for cement. The aggregates are delivered from the belts to their respective bins by means of rotating chutes and shuttle conveyors. The cement reaches the plant in box cars in bulk form and is elevated to its storage bin through a 5-inch pipe by either of two 40-hp. Fuller-Kinyon pumps which are actuated by compressed air. The bins and track hoppers combined provide storage for aggregates sufficient for 14 hours of plant operation and for cement for 3½ days. From these figures it is evident that the plant is dependent for continuity of operation upon the orderly functioning of the railroad which delivers the materials.

Water is obtained from the nearby Colorado River. It is first pumped to a 50-foot diameter Dorr clarifier located about a quarter-mile upstream from the plant. After all silt in excess of 500 parts to the million has settled out in this apparatus, the water flows by gravity to a 125,000-gallon tank situated adjacent to and higher than the mixing plant.

The plant contains probably the most efficient equipment ever devised for such an installation to determine accurately the proportions of the various ingredients of the concrete. It is designed to deliver cement within 1 per cent and aggregates within 2 per cent of the specified weights. As reported by J. Perry Yates, office engineer for Six Companies Inc., "it is required automatically to record, visibly and graphically, the time of measurement and the actual quantity of each of the aggregates, of cement, and of water for each batch; it must be readily adjustable for compensating for the varying amounts of moisture contained in the aggregates; and the water added must be weighed within 1 per cent. Also, the recorder must contain a device for indicating and registering the relative consistency of each batch."

Sand and the various sizes of gravel are each dropped through automatically controlled gates, in the bottoms of their respective bins, into a batch hopper or batcher. This batcher is mounted on the end of a dial scale which is connected, through levers and balancing weights to mercoid cut-off switches. By adjusting weights at the scales, the amount of aggregates to be included in each batch can be controlled.

There are two duplicate sets of batchers for the four mixers, each of which serves two mixers in alternation. Each set includes one batcher for sand, one for each of the three sizes of gravel, one for cobbles, one for water, and two for cement. Each batcher has its individual weighing, cut-off, and recorder beams and a full-capacity dial above the beam box.

The gates in the sand and aggregate bins are closed by compressed air. The following explanation of the manner in which each batch is accurately measured is from a description of the plant by W. R. Nelson, assistant engineer for the Bureau of Reclamation at the Hoover Dam: "The mercoid switches, installed on the weight end of the system, are operated by change in position of a glass capsule containing a globule of mercury. The electrical current is broken by two projections inserted at the end of the capsule and may be closed by lowering the end of the capsule containing the projections, thus allowing the electrical current to flow through the mercury globule. One of the controls is termed the 'main-flow cut-off', and the other the 'final-balance cut-off'. When the weight of the aggregates in the batcher reaches about 95 per cent of the predetermined batch, the swing of the beam, on which the controls are mounted, causes the mercury globule in the main-flow cut-off to move away from the two projections, thus breaking the electrical circuit and, by means of electromagnetic coils, closing the gate by compressed air. The electrical current operating through the mercury globule in the final-balance cut-off, by means of a small motor and an air valve, alternately quickly opens and closes the gate to allow small amounts of material to dribble into the batcher until the predetermined weight is acquired, at which time the adjusted position of the cut-off beam breaks the contact in the final-balance cut-off".

The cobble batcher is fed by a motor-

**Looking upstream from a point high up on the Arizona cliff, showing the mixing plant and a section of the canyon railroad over which materials move.**

driven apron conveyor which has a jog control to dribble the last portion and a final-weight cut-off. A special screw feeder, which also has a jog arrangement for adding the final portion, fills the cement batcher. The amount of water for each batch is regulated by means of a balanced valve which operates either automatically through solenoid control or by hand control, as desired.

The weighed batches of sand and gravel are conveyed to the mixer hopper by a collector belt; the cobble and cement batchers discharge directly into the mixer, and the water flows into it by gravity.

An operator on a stand located above and between each pair of mixers controls their operation. In front of him are signal lights showing when the batchers are filled and ready for discharging. To discharge the contents of the aggregate batchers, he pushes a button. Through the action of an electrical circuit, the doors of the batchers are opened in sequence, thereby producing a ribbon feed to the belt that conveys the materials to the mixer. The discharge gates on these batchers are unlatched by solenoids and are counterbalanced so as to close automatically when the batchers are empty. When these gates latch, they close an electrical circuit which again starts the flow of material into the batchers. Thus the filling of these batchers is entirely automatic and the whole cycle of operation is set in motion when the operator presses the button that empties them. The filling operations can be regulated by hand if so desired and the change to manual operation can be made quickly.

A valve at the operating stand actuates an air cylinder which controls the discharge of the cement batcher and another valve controls the flow of the water which is weighed and waiting. Because of this manually controlled discharge of these two ingredients, the operator can coördinate their introduction into the mixer with that of the aggregates as desired.

To start the mixing of a batch of concrete, the operator first turns on the water. After about 5 seconds he opens the gates of the mixer hopper, which contains the weighed aggregates, and the gates of the cement batcher. These actions permit all the ingredients of the concrete to enter the mixer together. After all are within the revolving drum, mixing continues for at least 2½ minutes, after which the concrete is discharged by tilting the drum.

The mixed concrete is delivered through a chute either to agitator drums mounted on trucks or to 2-cubic-yard buckets, both of which are in use for conveying it to points of operation. Later on it is planned to haul the concrete for the main dam structure in 8-cubic-yard buckets on railroad cars. With this in mind, railroad tracks have been laid; but as the same space is needed for truck haulage, the rails are set in a concrete roadway. The 8-cubic-yard buckets will be lifted from the cars and carried into position by cableway. They will be dumped by means of a trip line.

While one batch of concrete is being mixed, aggregates for the one to follow are being weighed and discharged into the mixer hopper and the cement and water batchers are being filled. The time required to empty a mixer of its charge and to introduce the ingredients for the next mix is never more than a minute. Thus far there has been no occasion to run the plant at more than 40 per cent of its capacity.

Moisture in the aggregates is compensated for by increasing the individual dry-batch weights by the computed water content, as determined from periodic tests, and by correspondingly reducing the weight of the water added to the mixer.

For each pair of mixers and the batchers that serve them there is an automatic recorder. It bears ten moving pens, arranged side by side, which transcribe a complete record upon a roll of graph paper which moves at constant speed. This record shows the weights of each of the several ingredients entering into each mix in turn and also registers the consistency of each batch and the length of time it was mixed. As the mixer revolves at constant speed and as the power required to operate it varies according to the volume of water in each mix, the power consumption indicates consistency with only small error. The chart furnishes a continuous graphic picture of what is happening in the plant and also provides a permanent record of each batch.

All batching equipment in the plant was designed and furnished by the C. S. Johnson Company, Champaign, Ill.

Equipment that insures uniformity of mixes. Top, left—Gravel batcher. Top, right—Cobble batcher. Center—Operator's station, with water batcher overhead. Bottom, left—Cement batcher. Bottom, right—Machine that records essential facts concerning every batch of concrete as it is mixed.

# Construction of the Hoover Dam

A truck carrying two 2-cubic-yard buckets of concrete into the portal of a diversion tunnel and over a bridge spanning the protecting cofferdam. Note the end of the concrete lining.

## *Lining of the Diversion Tunnels with Concrete*

C. H. VIVIAN

THE four diversion tunnels which will by-pass the Colorado River through the walls of Black Canyon while the Hoover Dam is being rooted in solid rock underlying the stream bed are being lined with concrete to insure their stability, strength, and carrying capacity while in service. Concreting operations have been in progress for more than eight months, and on November 1 were approximately 74 per cent completed. Just as the driving of these bores constitutes a new chapter in the annals of construction because of their magnitude, so does the application of concrete to their surfaces involve the doing of this particular thing on a bigger scale than it was ever done before. For the most part, standard practices and methods are being followed; but the element of hugeness, alone, is sufficient to make the work unique and to clothe it with interest. Nearly 2,000 tons of steel was used in making up the lining forms; and the cost of the equipment especially prepared for the work was approximately $325,000.

By way of refreshing the reader's memory, we should perhaps set down that there are four diversion tunnels—two on either side of the river. They range in lengths from 3,561 feet to 4,300 feet, and they total 15,909 feet, or about 3.1 miles. They are circular in cross section and were excavated to 56 feet in diameter. They are being lined with an average thickness of 3 feet of concrete, which will give them a finished section of 50 feet.

It is estimated that 394,000 cubic yards of concrete will be required for the lining operations. This would build about 135 miles of paved roadway 18 feet wide and 10 inches thick. Of this quantity approximately 77 per cent, or 303,000 cubic yards, will be paid for at the unit bid price of $11 per cubic yard. The remaining 91,000 cubic yards will have to be placed because of the overbreak of rock resulting from the impossibility of excavating evenly along the desired line, and is classed as nonpay yardage.

For purposes of concreting, the circular tunnel section is divided into three parts. The lower 74° comprises the invert, which is poured first. Above it, on either side, are side-wall portions of 88° each or 176° combined, and these are poured second. Last of all, the remaining roof or arch section of 110° is placed. Disregarding overbreak, it requires approximately 18.5 cubic yards of concrete for each linear foot of lining. This yardage is made up as follows: invert, 3.8; sides, 9; arch, 5.7.

Each of the component sections required to form a complete cylinder of lining is of the same length, so that a complete transverse construction joint is secured where such a cylinder abuts another one. The length of each section is 40 feet, except in those portions of the two outer tunnels which will later be used as spillways, where it is 26 feet 8 inches. Keyways are provided in all joints, both transverse and longitudinal, to knit the lining structure together as strongly as possible.

The concreting was started at the upstream

These pictures show progressive steps in the pouring of the invert or lower section of the concrete lining.

Left—Forms set up between longitudinal concrete risers placed at either side as bases for the gantry-crane rails.

The concrete poured by the gantry crane is mechanically troweled to the concrete surface curvature by a movable template.

A section of the invert has been completed and the template moved ahead. The crane is handling two concrete buckets.

Workmen are removing the forms from a finished section of the invert.

**Drilling holes in the arch for low-pressure grouting. Jumbos such as this, which were used for trimming the tunnel to line during excavating, have been adapted for grout-hole drilling. Sixteen N-75 drifter drills are mounted on this carriage.**

portals and carried progressively toward the downstream ends of the tunnels. A minor departure from this practice has been to start the invert pouring several hundred feet ahead and then work back toward the finished portion. All concrete is mixed at Lomix, the low-level mixing plant located a short distance upstream from the tunnel inlets on the Nevada side of the river. Two types of conveyances are used for moving the concrete from the plant to the pouring sites. Where the pouring must be done with the aid of gantry cranes, the material is transported on trucks in 2-cubic-yard bottom-dump steel buckets. Eleven trucks which had previously been used for hauling muck from the tunnels were converted from dump types to flat-bed bodies for this service. Each truck hauls two buckets at a time—a total of 4 cubic yards of concrete. Where the concrete can be poured directly into forms or into chutes leading to hoppers from which buckets can be filled as needed, the transportation is effected in Rex 4½-cubic-yard agitator mixers mounted on trucks. Fourteen such units were employed at the height of concreting operations. Trucks cross the river to and from the Arizona tunnels on a suspension bridge which was used to handle muck during the excavating of the bores. Concrete arch cofferdams have been built around each portal to keep water out of the tunnels during floods and are flanked on the river side by rock fills. To provide a means of passing over these barriers, wooden trestles were erected connecting with ramps leading down to the tunnel floor. During the lining of sections near the inlets, concrete was hauled to the portals in agitators and then chuted below to a steel hopper from which steel buckets were filled and trucked to points of use.

The first operation in preparing to pour the invert is the construction of two longitudinal concrete strips to serve as rail bases for the gantry crane. The inside face of each of these benches is 15 feet 9½ inches from the center line of the tunnel, and the top of this inside face is 24 inches from the point which will mark the finished surface of the lining. This interval, which corresponds to the minimum thickness of lining allowable under the specifications, is later filled and the track bases become integral parts of the lining. Timbers 6x12 inches are laid on each of these concrete strips, and 90-pound rails are spiked to them. The 10-ton electrically operated gantry crane has a maximum speed of 300 feet per minute along these rails. Its transverse bridge is equipped with two hooks, each with a lifting capacity of 5 tons and a hoisting speed of 100 feet per minute. Four such cranes have been in service—one in each tunnel.

After the rail foundations are in place, steel side forms for the invert are bolted to them. These forms are 2 feet high and are made up in sections of 10 feet. They are composed of 10-gauge steel plate supported by 2-inch angle irons at the top and by 2x3-inch stiffeners. Transverse forms of steel, shaped to conform to the radius of the invert section, are bolted to the longitudinal forms and also braced against the tunnel floor. They are 2 feet high, and the space between their lower edges and the tunnel floor is bulkheaded with 2-inch timbers. These transverse forms are spaced at intervals of either 26 feet 8 inches or 40 feet.

The invert surface is mechanically made to conform to the desired curvature by means of a template. The framework for this device consists of two transverse I-beams, spaced about 11 feet apart and connected at their ends by steel members. This frame runs along the tunnel on car wheels, the upper flanges of the longitudinal plates of the concrete invert forms being their tracks. The beams are shaped to conform to the curvature of the tunnel radius. Their lower inside flanges form tracks for wheels which support two cross members, each some 11 feet long and 4 feet wide and consisting of two decks. The bottom deck is a screed plate, which is also shaped to the curvature of the finished tunnel section. The two screeds are independently operated by hand winches on their upper decks. These winches control cables which pass over sheaves and fasten to the end pieces connecting the two I-beams. When concreting of an invert section starts, the two screeds are together

**Right—A section of the massive side-wall forms being assembled on the beach in Black Canyon. Above—Close view of a portion of a form in position for pouring concrete.**

in the center of the tunnel line. The gantry crane takes up two buckets of concrete from a nearby truck and moves them to the form. There they are dumped, one on either side of the screeds, by the manual operation of a hand wheel which controls their gates. The concrete thus deposited is puddled into place by workmen. When the concrete has been built up sufficiently high, the screeds are moved outward, toward the tunnel side walls, by means of the winches. By repetition of this procedure, until the edges of the form are reached, the required concave surface is obtained, the screeds acting in effect like huge inanimate trowels. The finishing touches are applied by workmen stationed on a movable timber platform suspended just above the concrete and supported between two curved I-beams mounted on flanged wheels which run on the same rails as the gantry crane. The screed framework can be jacked up sufficiently to obtain clearance and thus be moved manually to its next position. Moves of any considerable distance are made with the aid of the gantry crane. After a section of invert has been finished, a roadway for trucks is made by leveling it up with about 3 feet of sand.

The side-wall and arch sections of the lining are poured behind special forms. Because of the great size of the tunnels, standard forms could not be utilized. Those in use were designed by Six Companies Inc. and made up to their specifications by the Consolidated Steel Corporation of Los Angeles.

The pouring of the side-wall sections must necessarily follow some little time after the invert placing, as this lower portion of the lining is utilized not only as a haulageway for the concrete used later but also as a means of support for the forms which receive it. The first step is to pour two concrete strips 18 inches wide running lengthwise of the tunnel and resting on the invert. Ninety-pound rails are laid on these, each rail being 11½ feet from the center line of the tunnel. This 23-foot-gauge track constitutes the runway for the side-wall jumbo, a mammoth structural-steel framework 80 feet long and 50 feet high and weighing 385 tons. It supports the wall forms composed of ¼-inch steel plate. To render it usable on curves, where the two opposing walls are of different lengths, the jumbo is made up in five sections, three of 20 feet and two of 10 feet. Wooden panels are built up between the steel panels where a gap results in rounding a curve.

The jumbo is completely equipped for handling concrete and pouring it where desired. At its top is a 5-ton, electrically operated bridge crane which can travel longitudinally at the rate of 300 feet per minute on 50-pound rails. Its transverse traveler can move at a speed of 125 feet per minute, and its two steel hooks have hoisting speeds of 100 feet per minute. Concrete is elevated in 2-cubic-yard buckets and poured into chutes of rectangular section which lead to the faces of the forms. These chutes are of ¼-inch steel, reinforced by 2-inch angle irons. They are 30 inches wide, 12 inches deep, and from 8 to 16 feet long, and there are six longitudinal rows of them spaced from 4 to 6 feet vertically at the form faces. The openings in the form faces to which the chutes lead are fitted with 12x24-inch steel doors or gates which can be manually closed and bolted.

In addition to the six rows of simple chutes there is a row at the top of each form of what are designated as "coffin" or "bathtub" chutes. Each is hinged at the form line, which permits lowering the end toward the center of the tunnel while the depressed chute is being filled and then raising the loaded end to give the chute sufficient slope so that the concrete will flow to the form by gravity. These chutes are raised and lowered by a cable which passes over sheaves and which is operated by an Ingersoll-Rand Size HU "Utility" hoist, two of which are installed at the base of the jumbo. The upper 4 feet of the side-wall sections is poured by this means. A series of screw jacks and ratchets is provided for adjusting the wall forms for pouring, for releasing them from the finished concrete faces, and for providing for the distribution of the hydrostatic pressure of green concrete.

Pouring is, of course, started through the lowest row of chutes and carried progressively upward. The two opposite walls of a section are poured simultaneously, and individual sections of either 26⅔ or 40 feet in length are completed to the top of the forms before adjoining ones are started. To provide these shorter sections while using an 80-foot form, a timber bulkhead is erected at the proper

Left—Side-wall forms in position, with two buckets of concrete being elevated for pouring into the chutes. Above—Side-wall forms being made ready for use.

point. This is framed so as to leave a keyway in the end of the section 10 inches wide and 1½ inches deep. A round iron bar, 3 inches in diameter, is secured over the entrance to each chute. In pouring, the crane operator maneuvers each bucket of concrete into such a position that hooks on the bucket gate are directly above this bar. Lowering the bucket trips open the gate, and the concrete pours into the chute. The bucket is then disengaged from the bar, lowered to the truck that brought it, and the second bucket is raised from the truck bed and poured in a similar manner into the corresponding chute on the opposite side of the form. Behind each form, in the 3-foot space between the steel plate and the rock wall of the tunnel, are from five to seven men, who puddle the concrete into place, and an inspector. When a section has been built up to the level of the row of chutes through which the pouring has been done, the gates are closed over those openings and bolted securely, thus making them integral parts of the form for subsequent pouring. Delivery of concrete is then started through the next higher set of chutes, and this procedure is repeated until the section has been completed to the top of the form. The final operation is to form a keyway approximately 10 inches wide and 2 inches deep on top of the section, where the arch will later rest.

It requires about 50 hours of elapsed time to complete an 80-foot section of side wall, which includes both sides of the tunnel, and to set up for the next one. The pouring itself consumes about 24 hours. The forms are left in place ten hours longer. The timber bulkhead at the end of the form, which does not bear against a previously formed section, is then removed; jacks and ratchets are loosened; and the jumbo is advanced to its next position by means of a block and tackle secured to the rails ahead and operated by the compressed-air hoists at the base of the form structure.

The arch or top section of the tunnel is lined pneumatically. The structural-steel jumbo that supports the forms and carries all essential apparatus is really a 3-part structure, all the units of which are mounted on flanged wheels that run on the steel rails used by the side-wall jumbo. The three parts of this jumbo are: an arch-form support, a concrete-gun carriage, and a pipe carriage. The general features of the arch-form support, and the manner in which it holds the steel plate or skin on the top of which the concrete is placed, are clearly shown in one of the accompanying pictures. The forms are moved into position for pouring or withdrawn from the finished concrete face by means of screw jacks. The framework for the gun carriage is approximately 46 feet high and 45 feet long and it is run along its track by a 25-hp. motor at a maximum speed of 100 feet per minute when going forward and 20 feet per minute in the reverse direction. On each side, at the base, is mounted a 2-cubic-yard Hackley concrete placement gun, with hopper. Overhead, near the top of the framework, are two air receivers, one on either side, that supply the surge of compressed air necessary for discharging each batch of concrete. Air is piped into the tunnels from the established distribution systems.

Next to the gun carriage is the pipe carriage which supports the 6-inch wrought-iron pipe through which the concrete is conveyed to the forms. The traveler is stationed between the arch-form jumbo and the pipe carriage.

An electric double-drum hoist on the gun carriage lifts a 4½-cubic-yard agitator of concrete from a truck and discharges it into the hopper of either of the guns. The batch is then shot upward and forward through the delivery pipe and its rubber-hose connections to the arch form ahead. Placing is started at the end of the form farthest from the concrete guns; and the equipment is moved rearward as the work progresses. Construction joints with keyways are secured at intervals of 26⅔ or 40 feet by placing bulkheads.

The concrete was first cured by sprinkling the surfaces with jet sprays as soon as the forms were removed and keeping them continuously wet for fourteen days. River water was pumped from sumps near the tunnel portals by Cameron Type HV centrifugal pumps. Since May all curing has been done by applying a Hunt process asphaltic paving coat. The spraying is done with compressed air.

Specifications for concrete in the tunnel lining are very rigid, and insistence is placed on each batch being as nearly identical as it is possible to make it with the others of a like mix. The Bureau of Reclamation main-

Arch jumbo under construction. Human ants may be discerned at several points.

Arch form being set up at a tunnel portal. Invert and side-wall lining sections are in place.

Left—Concrete-gun carriage used in placing the arch section of the lining.

Below—Filling a bucket with concrete from an agitator truck. Side-wall pouring chutes are at either side.

Right—Interior of a tunnel with invert and side-wall sections of the lining completed.

tains a laboratory adjacent to the mixing plant, and a large staff of technicians and inspectors is in constant attendance. Cylinders are made regularly throughout the day, and are tested in the usual manner.

Concrete for invert and side-wall sections contains 1 part cement. 2.1 parts sand, and 4.7 parts gravel. A 4-cubic-yard batch weighs 16,600 pounds, distributed as follows: cement, 2,000 pounds; sand, 4,200 pounds; gravel, 9,400 pounds; and water, 1,000 pounds. Thirty-six per cent of the gravel, or 3,380 pounds, is of the coarse grade, which ranges from 1½ to 3 inches; 32 per cent, or 3,010 pounds, is of the intermediate size, ¾ inch to 1½ inches; and 32 per cent, or 3,010 pounds, is of the fine size, ¼ to ¾ inch.

The allowable slump in the invert and side-wall concrete is from 3 to 4 inches. The slump of concrete, it may be explained, is the measure of its consistency. The standard method of determining slump consists of forming a cone—12 inches high, 8 inches in diameter at the base, and 4 inches at the top—by filling a mold with four layers of concrete and tamping each layer 25 times with a ⅝-inch rod which is bullet pointed at the end. The mold is removed immediately after being filled. After the concrete has settled, the amount of sag or slump from its original height is measured.

The largest size of gravel, 1½ to 3 inches, is omitted from concrete for the arch section, for which the following ratio is specified: 1 part cement, 2.5 parts sand, and 4.3 parts gravel. By weight, a 4-cubic-yard batch contains 2,000 pounds of cement, 5,000 pounds of sand, and 8,600 pounds of gravel. The gravel is 60 per cent of intermediate size and 40 per cent of fine size. Sufficient water is added to the concrete to give it a 6- or 7-inch slump.

After the tunnels have served their purpose—that of diverting the water around the dam site, they will be used for other purposes. The two farthest from the river, Nos. 1 and 4, will become spillway outlets. To accomplish this, each of the tunnels will be plugged with reinforced concrete about midway of its length for a distance of 396 feet. This plug will be just upstream from the entrance of each of the inclined spillway tunnels, which will extend to the surface and emerge at the approximate level of the top of the dam.

Each of the two inside tunnels, Nos. 2 and 3, will become a penstock tunnel for a portion of its length and will house a 30-foot-diameter steel penstock pipe. To convert these bores for this service, upper and lower concrete plugs will be placed in each of them. In order that irrigation water may be passed to the Blythe, Yuma, and Imperial valleys during the period that the reservoir is being filled to the level of the intake towers—approximately 300 feet, it is expected that the heavy slide gates will be operated through these upstream plugs. Trash-rack structures will be built at their intake portals. Each of the downstream plugs will later contain six 72-inch needle valves, which will be connected with the penstocks, and the upstream gates will be plugged. At the downstream portals, 50x35-foot stoney gates will be built.

The two outside, or spillway, tunnels will be open at their lower ends, but their intake portals will be closed off with steel bulkhead gates, each of about 1,100 tons weight.

When the tunnels were driven, excavations were also made for the tunnel plugs, which are approximately 12 to 16 feet larger in diameter than the 56-foot tunnel bores. They are irregular in shape and resemble in cross section three wedges linked together endwise. The specifications provide that the tunnel-plug excavations must be lined with 3 feet

**Chuting concrete down into a tunnel from the top of a cofferdam at one of the portals.**

of concrete, corresponding to the rest of the tunnel lining, but that they must likewise present smooth surfaces continuous with the standard 50-foot lined sections during the period of diversion. Six Companies Inc. were given the alternative of filling in the annular spaces with concrete or of erecting timber falsework in them. As it would have been necessary to remove the concrete in order to install the plugs, the latter method was chosen; and timber panels have been constructed which will present smooth center surfaces to the water and which are fastened to the concrete lining with anchor bolts in such a manner that they can readily be taken down.

These forms are very intricate, and the designing of them called for the exercise of much engineering resource and ingenuity. A large concrete surface was poured to serve as a drawing board on which to lay out the measurements of the hundreds of different shapes of panels required. A small sawmill and a crew of 25 carpenters were engaged for several months in making them up. After being cut to specific sizes and shapes, all timber sections were numbered so that they might be assembled in proper order and position in the tunnels. They were then stored in rows on the desert to await use. The average size of the timber pads is about 20 feet in length and from 4 to 6 feet in their other dimension. As many as 425 different shapes and sizes were required for one structure. When one of these forms is being installed, the steel form for the arch or side-wall section is moved into position. The timber pads are then built up around it; bolted in four directions; and anchored in the concrete lining of the larger section, as the concrete is poured by means of chutes which pass through both the steel and the wood forms.

Specifications call for both low-pressure and high-pressure grouting of the tunnels. Low-pressure grouting, which is done only in the arch section, is largely for the purpose of filling voids which may occur between the lining and the rock, and to compensate for any shrinkage which may have occurred. Holes for introduction of the grout are drilled with Ingersoll-Rand N-75 drifter drills, sixteen of which are mounted on a modification of the steel framework which was used in trimming the tunnels to line. Three holes, each 2 inches in diameter, are driven in the arch section—one in the center, and two at 45° angles from the perpendicular, as well as special holes to overbreak points. A set of these holes is drilled every 20 feet along the tunnel line.

Grout, consisting of neat cement and water, is applied at from 50 to 100 pounds pressure by means of a duplex, air-driven, piston grout pump which is mounted on a truck together with a mixer. The truck also hauls a trailer loaded with bags of cement. Water is obtained from pipe connections on the line which extends throughout the length of the tunnels.

High-pressure grouting will be done in certain portions of the tunnels at the direction of Bureau of Reclamation engineers. It is expected that the holes for this purpose will extend 20 to 30 feet into the rock and that grout will be forced into them under a pressure of from 300 to 500 pounds. It is believed that not much grout will be required owing to the dense, compact character of the rock and the absence of cleavage and shear planes. These high-pressure holes will be staggered radially, eight in a ring, around the tunnel, and grout will be applied merely as an added precautionary measure to strengthen the lining in zones which will be subjected to particularly heavy stresses and high velocities when the tunnels are in service.

Lining operations were started in Tunnel No. 3 on March 16, 1932, and in the other

**Placing temporary timber lining, which conforms to the 50-foot finished concrete lining of other sections, in an enlarged part of the tunnel that will later be plugged with concrete. This view shows details of the side-wall lining forms.**

bores, as follows: No. 2, March 29; No. 4, April 7; and No. 1, August 3. The following table shows the quantity of concrete placed, in cubic yards, by months:

| | PAY YARDAGE | | | |
|---|---|---|---|---|
| | Invert | Side | Arch | Total |
| March.... | 1,064 | .... | .... | 1,064 |
| April...... | 8,664 | 405 | .... | 9,069 |
| May...... | 4,161 | 15,210 | .... | 19,371 |
| June...... | 11,518 | 16,816 | 4,102 | 32,436 |
| July...... | 14,440 | 27,243 | 10,966 | 52,649 |
| August.... | 8,409 | 27,461 | 14,873 | 50,743 |
| September. | 1,280 | 14,269 | 17,803 | 33,352 |
| October... | 3,210 | 13,796 | 9,655 | 26,661 |
| Total..... | 52,746 | 115,200 | 57,399 | 225,345 |
| | GROSS YARDAGE PLACED | | | |
| | 77,771 | 143,862 | 69,369 | 291,002 |

During August, a typical month, the production of the concrete mixing plant averaged 171 four-cubic-yard batches per shift, or 2,052 cubic yards every 24 hours. The two best days of tunnel lining thus far were June 13 and 14, when 2,646 and 2,750 cubic yards, respectively, were placed during a 24-hour period. The average of concrete poured per day has been about 1,400 cubic yards, of which 1,080 was pay yardage.

The following tabulation shows the percentages of the various sections of lining that had been placed up to November 1.

| Tunnel | Invert | Sides | Arch |
|---|---|---|---|
| No. 1 | 55% | 32% | 5% |
| No. 2 | 100 | 92 | 76 |
| No. 3 | 100 | 100 | 86 |
| No. 4 | 100 | 100 | 96 |

About 900 men, working 300 per shift, were engaged in various capacities in connection with the tunnel lining when work was at its height. On a typical day, 916 men were occupied as follows: cleaning up invert and side walls, 142; grouting, 77; placing concrete, 199; trucking concrete, 53; mixing plant and water plant, 43; erecting wood forms, 70; erecting tunnel-plug forms, 63; building wood forms in shop, 27; moving invert, side, and arch forms, 112; trucking materials, 8; miscellaneous, 122.

Tunnels Nos. 3 and 4, both on the Arizona side of the river, were completed early in November, and on November 13 and 14 the stream was diverted through them. The first water entered Tunnel No. 4 after a blast had cleared a channel through the cofferdam in front of the portal. A diversion dam was then hastily constructed across the canyon just downstream from the tunnel intakes. This was accomplished by dumping trucks of rock from a trestle bridge which spans the river there. It required 30 hours, during which period an average of four truckloads a minute was deposited, to complete this barrier. A similar dam was thrown across the canyon just upstream from the tunnel outlets to keep the water discharging from the tunnels from backing up in the stream bed. These rock dams are temporary, as cofferdams, higher and tighter, will be constructed to uncover approximately 3,800 feet of the river bed until excavations have been made and the dam footings placed.

The diversion of the river ends the preliminary or preparatory stage of construction. The work at the dam site proper can now proceed unhampered. If all goes according to schedule, the excavations will be ready to receive the first concrete next summer.

HIGH scalers at their perilous work of stripping loose and projecting rocks from the walls of Black Canyon. Suspended by cables, these workmen accomplish their task with "Jackhamers" and duralumin pinch bars.

# Construction of the Hoover Dam

*A Resumé of Current Activities and an Account of the Building of the Cofferdams**

WESLEY R. NELSON†

**Hundreds of tons of rock were dumped from a trestle bridge in November, 1932 to form a temporary diversion dam.**

A BLAST at the inlet of Diversion Tunnel No. 4, half an hour before noon on November 13, 1932, turned a portion of the turbid Colorado River from its natural course into a man-made by-pass through one of the walls of Black Canyon and signified the completion of the first stage of the construction program which will ultimately conserve the wasted energies of the stream and employ them in the development of the Southwest. This initial diversion of the river around the site where Hoover Dam will be reared marked the successful termination of most of the preparatory work—the principal exceptions being the cofferdams upstream and downstream from the dam site. It likewise indicated the commencement of many phases of the permanent structures.

Previous articles in this series of booklets have described the adequate railroad and highway systems that connect the working zone with trunk lines and related how the modern construction camp of Boulder City was developed from waste desert land to a community of 1,000 homes and 5,000 population within scarcely a year. Others have portrayed the adequacy and magnitude of the plants and equipment installed by Six Companies Inc. to expedite construction and told of the revolutionary measures by which 1,500,000 cubic yards of rock was removed from three miles of 56-foot-diameter tunnels in one year and a 3-foot lining of concrete placed in them in a similar period by the pouring of approximately 370,000 cubic yards of concrete.

The same dispatch and care with which the preliminary operations were completed is in evidence today in the work in progress. The 600-foot-long inclined tunnel from the Nevada spillway channel to Diversion Tunnel No. 1 has been enlarged from a 7x14-foot top heading to an opening with an average diameter of 56 feet in about 125 days, during which time nearly 50,000 cubic yards of rock has been moved. Approximately 80 per cent of the required excavation has been shot out of the canyon walls in the cuts for the 30-story, 75-foot-average-diameter, concrete intake towers through which water will flow from the reservoir to outlet works or to turbines in the power plant.

Loose and projecting rocks are being pried and blasted from the canyon walls as a precautionary measure—rocks that might otherwise fall on structures while being built or after construction. This work is being done by high scalers, suspended by long ropes, who dislodge the loose bowlders with duralumin pinch bars, remove the fine material with compressed-air or water jets, and drill and blast away the more solid and protruding rock until the cliff presents a relatively smooth and even slope.

Other work in progress includes the driving of penstock tunnels, of an inclined raise to the upstream intake-tower site, and of a 26x43-foot construction adit—all from the Nevada diversion tunnel nearer the river. A shelf has been cut on each canyon wall for outlet works; tunnels are being driven inward from these; and from points near the outlet works have been started 26x43-foot construction adits which will lead to the pressure tunnels. From these connections the pressure tunnels will be driven upstream toward the sites of the downstream intake towers. Stoney gates, 50x35 feet in vertical-plane dimensions, are under erection at the outlets of the two diversion tunnels closer to the river, and 50x50-foot steel bulkheads are being placed at the inlets of the tunnels farther from the river. Each huge bulkhead, complete with its steel frame and cylinders, weighs over 3,000,000 pounds, and required 42 railroad cars for transportation to the proj-

†Assistant Engineer, U. S. Bureau of Reclamation, Boulder Canyon Project.

**Thirteenth of a series of articles on the Colorado River and the building of Hoover Dam.*

**Right—Half the stream bed inclosed by a dike to permit an early start on the upstream cofferdam.**

**Center—Inlet portals of the Arizona tunnels after the initial diversion of the river.**

**Bottom—Looking upstream after the diversion at the area in which Hoover Dam will take form.**

ect. Eminently important at the present time, owing to the menace of spring floods, is the construction of two diversion cofferdams that will maintain the flow of the river through the diversion tunnels and provide a dry site on which the main structure of Hoover Dam will be built.

As is shown in accompanying illustrations, the upper cofferdam is an earth fill. The upstream face is protected by 6-inch reinforced-concrete paving laid on a 3-foot thickness of rock blanket, and the downstream face is covered by a heavy rock fill. The slope is 3 to 1 on the upstream face and 4 to 1 on the downstream face. A cut-off of steel sheet piling at the upstream toe rests on bedrock and is joined to the face paving by a concrete wall and rubber seal. A puddled-clay fill and a rock blanket protect the connection.

A rubber seal is also placed at each junction of the face paving and canyon wall and extends the full width of the paving from the steel piling to the crest of the dam. A ledge for this connection is cut in each canyon wall contiguous to the face paving and on the ledge is built a concrete curb with a 4-inch space between it and the paving. A 36-inch width of tire fabric, $\frac{1}{4}$ inch thick, is folded down into this space—the edges of the fabric being fastened to both curb and paving. Thus any settlement of the dam or other movement in the face paving will not break the seals between the dam and the canyon walls.

Three reinforced-concrete percolation stops, resembling thin counterforts, are poured near the center of the dam where it contacts with the canyon walls. Channels 3 feet deep are cut into the cliffs from base to crest, and concrete is poured to form piers of 2-foot basal thickness, approximately 8 feet across the sides, and 98 feet high. The percolation stops extend at least 5 feet into the dam fill and prevent the free flow of water along the plane

Left—Site of the upper cofferdam, with the contour of the structure painted on the canyon wall.

Below—A view upstream on January 6, 1933, showing water emerging from the Arizona diversion tunnels.

between the cliffs and the dam.

The upper cofferdam is 98 feet high, 480 feet long, and 750 feet thick at the base. More than 500,000 cubic yards of earth and gravel, 151,000 cubic yards of rock, and 3,500 cubic yards of concrete were required for its construction. The face paving covers an area of more than three acres and contains ⅝-inch reinforcing material, on 15-inch centers both ways, comprising nearly four miles of steel bars weighing 215,000 pounds. The dam was given these dimensions and features to turn a river flow of more than 200,000 second-feet through the diversion tunnels and to reduce the percolation through the dam to a minimum.

Specifications 519—governing the contract for the construction of Hoover Dam and the power plant and appurtenant works—stipulated for both cofferdams that "the earth fill portions of the cofferdams shall consist of a mixture similar to the natural mixture of silt, sand, and gravel in Hemenway Wash." Other requirements that influenced the contractors' procedure of construction appear in the specifications, as follows:

"No stones having maximum dimensions of more than 9 inches shall be placed in the earth fill embankments.

"The mixture of silt, sand, and gravel shall be placed in the embankment in approximately horizontal layers not more than 12 inches in thickness after rolling. All materials shall be uniformly moistened on the embankment. . . . to such a degree that. . . . the maximum compactness of embankment will be assured after rolling as specified. The roller used. . . . shall be of the 'Rohl' type, having cast iron ball feet equally spaced over its cylindrical surface and its weight shall be not less than 1,000 pounds per linear foot of width of tread.

"The largest rock permissible in the rock fills shall be not more than 1 cubic yard in volume. . . . It is contemplated that material direct from any of the required rock excavation will be suitable.

"The rock in the rock embankments need not be especially compacted but shall be built up by dumping and roughly leveling. . . . such as to insure that the completed fill will be stable, without tendency to slide, and that there will be no unfilled spaces within the fill.

"The upstream slope of the upstream cofferdam shall be covered with a 3-foot layer of . . . . sound, durable rock fragments or bowlders the maximum dimensions of which shall be not less than 4 inches nor more than 30 inches. . . . The rock shall be evenly distributed or leveled to a roughly uniform layer of the required thickness, on which the concrete paving is to be placed. . . . The method used in placing the rock blanket shall be at the option of the contractor, provided that the method shall secure the required compactness necessary as a foundation for the concrete paving. . . . The contractor shall place any gravel or small rock fragments. . . . necessary to fill the voids in the upper surface of the rock blanket and furnish a suitable surface on which to lay the concrete paving.

"The concrete paving shall be laid in strips not over 16 feet in width running up and down the slope of the cofferdam, and construction joints, across which the reinforcement bars shall be continuous, shall be provided between the strips.

"The rock barrier below the downstream cofferdam shall consist of hard, dense, and durable rock, equivalent in this respect. . . . to the best rock for resisting wear or erosion that exists in the vicinity of the work, and that may be readily obtained from nearby borrow pits, quarries, or from required excavation. There will be no limit to the size of rock permitted in the rock barrier. . . . The rock may be loosely dumped but in such order and in such a manner as will. . . . insure that the completed fill be stable, without tendency to slide, and so that there shall be no unreasonably large unfilled spaces within the fill.

"Measurements, for payment, of materials in the cofferdams. . . . and rock barrier will be made in place."

Early in September, 1932, an end-dump fill, using muck from the Nevada spillway inclined tunnel, was started from the Nevada side of the canyon and downstream from the inlet portals of the Nevada diversion tunnels. In succeeding weeks the fill was extended to the center of the canyon, thence downstream and back to the Nevada side, thereby inclosing the Nevada half of the upper cofferdam site and crowding the river into the Arizona side of the channel. The inclosed area was pumped dry; a 5-cubic-yard electric dragline was brought down from the Arizona gravel deposit; three 3½-cubic-yard electric

**Blasting away rock for the Nevada abutment of Hoover Dam. The road will cross the crest of the structure.**

**This picture and the opposite one show the finishing operations on the upstream face of the upper cofferdam. The photograph was made January 5, 1933.**

shovels and a fleet of trucks were moved to the site; and excavation of the silt and sand was started to uncover a consolidated formation suitable for the cofferdam foundation. This formation was discovered at elevation 622, at a point 18 feet below the river bed and 98 feet below the dam crest.

Desirable material for the earth fill had been located by test holes in Hemenway Wash. Pits were opened with 3½-cubic-yard electric shovels; railroad track was laid to the pits from the Black Canyon line of Six Companies Inc.; and later the shovels were used for loading the material into 30-cubic-yard side-dump cars for transportation to an unloading site near the cofferdam.

By the time the diversion tunnels were being excavated, a railroad had been built into Black Canyon as far as No. 1 Compressor Plant, half a mile upstream from the cofferdam site on the Nevada shore, and two tunnels had been driven on the continuation of the line—one 1,100 feet in length extending downstream from the compressor plant, and the other 900 feet long extending downstream from the inlets of the Nevada diversion tunnels. The grade elevation of the railroad and the tunnels is 720, the same as the crest of the cofferdam.

A fill was then built along the river's edge from the compressor plant downstream to a location just upstream from the inlets of the Nevada diversion tunnels; a single-track line was laid on the fill outside the 1,100-foot tunnel; and double tracks were run from near the lower portal of this tunnel to the cofferdam. The track closer to the river was extended by a pile trestle bridge to a point downstream from the diversion-tunnel inlets, providing a structure for unloading fill material brought from Hemenway Wash. A loading platform was constructed on the canyon-wall side of the other track for removing the silt and sand transported from the cofferdam excavation by 8- and 10-yard trucks. This rejected material was used for widening the railroad bench along the river or was unloaded at a dump ground near the mouth of Hemenway Wash.

The material from Hemenway Wash was dumped from the trestle by side-dump cars, loaded into trucks by two 3½-cubic-yard electric shovels, and moved to the cofferdam fill. There it was dumped by the trucks; spread by bulldozers mounted on 60-hp. Caterpillar tractors; the correct amount of water sprinkled on the material; and the mixture rolled by 6-ton Rohl type rollers pulled by 30-hp. Caterpillar tractors. When excavating for the cofferdam was in progress, the trucks that hauled the fill material also proceeded to the site of the excavation and there received the loads of silt and sand which they took to the loading platform and dumped into the waiting railroad cars.

A pile trestle bridge had been built across the river downstream from the inlets of the diversion tunnels, and as soon as the blast removed the barrier in front of Tunnel No. 4, allowing water to flow through it, trucks commenced hauling rock and dumping it into the river on both sides of the bridge. Within 30 hours after the initial diversion, the temporary dam thus constructed had risen high enough to cause all except a small amount of water seeping through the hastily constructed fill to flow through the Arizona diversion tunnels. To cut off the seepage and to prevent inundation of the cofferdam site by a possible river flood, the temporary dam was heightened and widened, using muck from the cofferdam excavation.

Soon after this temporary dam was completed above the upper cofferdam, a similar

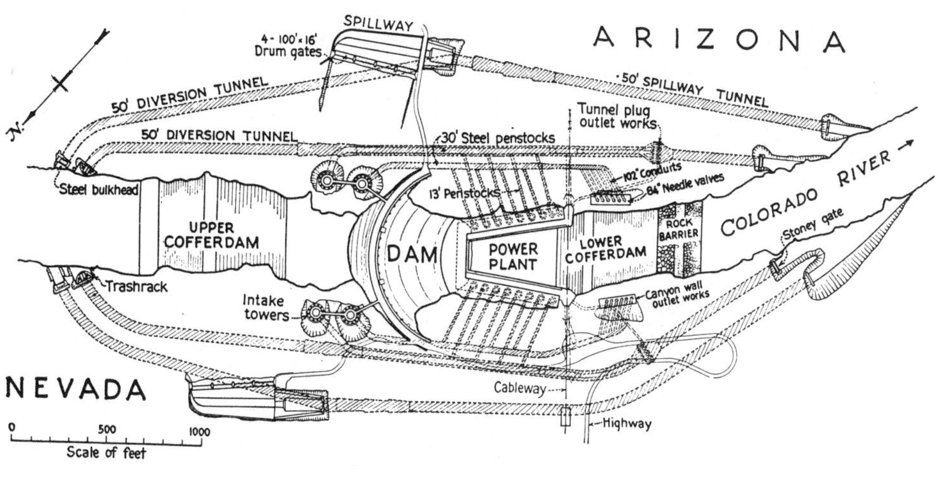


**Plan drawing of the principal structures involved in the rearing of Hoover Dam.**

Concrete is being laid in longitudinal sections. At the right a dragline is handling a hammer which is driving steel sheet piling with compressed air.

To compact it thoroughly, the earth fill in the cofferdam was kneaded with 6-ton sheep's-foot rollers.

fill was placed immediately upstream from the outlets of the diversion tunnels. The water in the river channel thus inclosed was pumped out, and excavations for the upper cofferdam were extended to the Arizona wall. The work was started on September 25 and finished on December 5, during which period 212,872 cubic yards of sand and gravel was removed at the average rate of 5,200 cubic yards a day.

Because of the existence of a number of large bowlders along the line of the upstream toe of the cofferdam on the Nevada side, the steel sheet piling was placed on a concrete footing, set on bedrock, for a distance of 120 feet out from the canyon wall. Along the remainder of the cut-off line the piling was driven by a McKiernan-Terry No. 9-B-2 hammer operated with compressed air. Ordinarily, the hammer was spotted and held in driving position by the 5-cubic-yard dragline. The piles are 16 inches wide, from 40 to 55 feet long, and are of the arch-wedge type, which has a continuous interlock rolled integral with the pile for its entire length. The 80-foot boom of the dragline was used to advantage in getting the piling in position.

The percolation stops were formed and poured as the fill progressed upward. The concrete required was hauled from the mixing plant by 10-ton trucks in 4-cubic-yard agitators; and the 5-cubic-yard dragline was used as a crane in lifting the agitators from the trucks to the pouring site.

The rock blanket on the upstream face and the rock fill on the downstream face were placed after the earth fill was practically completed. The material was secured principally from rock shot down from the canyon walls during the stripping operations or from the excavations for the intake towers.

The face paving was poured in 16-foot sections running from the concrete curb at the sheet piling to the crest of the dam. Paving was started on December 20, 1932, and at the end of January, 1933, was approximately 50 per cent completed. The ledges were cut in the canyon cliffs and the concrete curbs poured after the earth fill was finished.

Work on the earth fill was begun on October 31, 1932, and finished January 1, 1933. It contains approximately 510,000 cubic yards of material, of which 420,000 cubic yards was placed in December. On many days in that month as many as 4,000 truck loads, or 18,000 cubic yards of material, were deposited in the fill, using three $3\frac{1}{2}$-cubic-yard electric shovels at the pits and two or three shovels of the same capacity for loading the fleet of 35 trucks. For hours at a stretch, each shovel was loading, swinging, and dumping a shovelful in less than 30 seconds.

Construction of the lower cofferdam and rock barrier has been delayed by reason of the scaling of the canyon walls, making the site for the time being impractical of access. The rock barrier will consist of a fill of 100,000 cubic yards of rock. It will have a height of 54 feet, a length of 380 feet, a thickness of 210 feet at the base and of 50 feet at the crest, and slopes of $1\frac{1}{2}$ to 1 on each side. Its foundation is at approximate elevation 629, or 11 feet below the river bed.

The earth fill upstream from the barrier, although smaller, is similar in design to the upper cofferdam and calls for essentially the same procedure of construction. There are, however, no sheet-piling cut-off walls and no downstream-face covering or seals at junctions with the canyon walls; but, as in the case of the upper dam, two percolation stops will be provided on each side, and the upstream face will be covered with a heavy rock fill. It is estimated that the dam will contain 230,000 cubic yards of earth and 63,000 cubic yards of rock. It will be 360 feet long, 66 feet high, and range in thickness from 500 feet at the base to 50 feet at the crest. It will rise from a foundation elevation of approximately 624. The downstream slope is to be 5 to 1, and the upstream slope 2 to 1.

By February 1, 1933, the rock-barrier site had been excavated and 80,000 cubic yards of rock deposited for its construction. Very little work had been done up to that time on the lower cofferdam, but plans had been made to secure the earth fill from Hemenway Wash. This material will be hauled by railroad through the two long tunnels, dumped from a trestle near the main dam site, and loaded by power shovels into trucks for transportation to the cofferdam site.

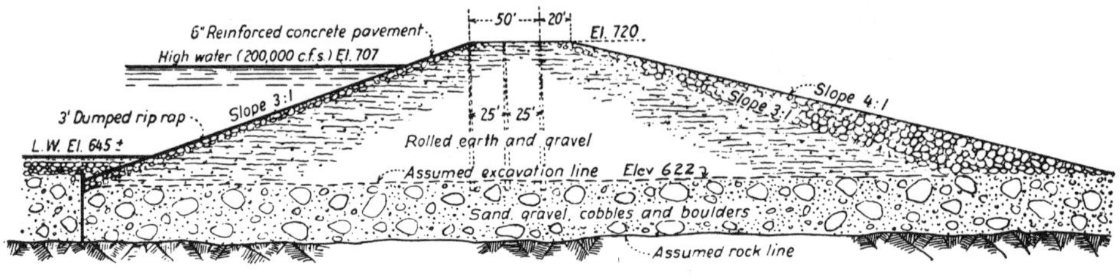


Cross section through the upper cofferdam showing the character of construction.

Left—Starting a 21-foot penstock tunnel from the interior of one of the 56-foot diversion bores.

Below—Outlet tunnels and (at the right) a construction adit being started from a shelf on the Nevada cliff.

Both cofferdams and all four diversion tunnels are expected to be completed before the high-water period of the coming spring and summer. Advantage was taken of the low flow of the Colorado River in constructing the upper cofferdam, the discharge during the months of November and December never having gone above 7,500 cubic feet per second. Flash floods of 50,000 cubic feet per second may occur in any month of the year, as happened in February and September of 1932. The greatest flow of the river is expected in May or June, when the discharge may exceed 200,000 cubic feet per second.

Eventually, when the bulkhead gates are closed at the inlets of the diversion tunnels, and when the flow of the river is controlled by the slide gates installed in the upstream plugs of the other two tunnels, the upper cofferdam will be overtopped by the waters of the Colorado and inundated by the reservoir back of Hoover Dam. The lower cofferdam and rock barrier will be removed from the river channel at about the same time, probably in 1935. The removal of these downstream fills is necessary to clear the tailrace for the efficient operation of the power plant.

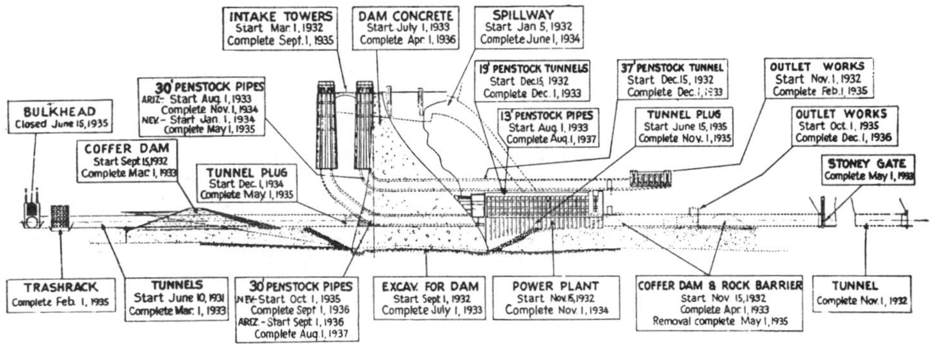


A recent Bureau of Reclamation diagram which shows the program of construction.

Erection view of the head tower of cableway 8, showing the concrete counterweight and the tracks on which the tower moves laterally. Note super-elevation of the tracks on the left to resist side pull.

# Construction of the Hoover Dam

*Description of the Aerial Cableways for the Transportation of Men, Materials, and Machinery.**

**WESLEY R. NELSON.†**

The head tower of cableway 7 under construction. In the background is the Arizona side of the canyon. The tail towers of cableways 5 and 6 can be discerned on the fill in the upper left-center of the picture.

AMONG the interesting and useful equipment for handling the immense quantities of construction materials for Hoover Dam are the ten cableways that span the gorge of Black Canyon at strategic positions in the mile-long zone of construction activities. Their facility for handling large machinery quickly and easily, and their rapid dispatch of men and materials to locations on the cliffs, have aided greatly in advancing the program of construction far ahead of schedule, and will insure rapid progress for the remainder of the work.

These aerial transports fascinate the observer as they move heavy loads with no apparent effort. A steel hook swoops down, virtually from the sky; cable slings are quickly adjusted about a caterpillar tractor, a 12-yard truck, or half of a 3-yard power shovel; and the object is lifted smoothly from the ground, carried easily upward and outward above the yawning chasm, and then gently lowered to the edge of the river or to some ledge high on the canyon wall.

At another place, on the Nevada rim, day-shift workers are disembarking from a cableway skip, and members of the swing shift are taking their places on the railed timber platform that hangs suspended from the 8-wheel track carriage above. The skip tender waves a signal to the hoistman, the skip rises from the ledge, hesitates as it nears the track cable, then moves at a speed of 1,200 feet per minute out over the canyon. At intervals during the trip, a projection on the button cable overhead picks up a slack-rope carrier from the group attached to the carriage, an expedient to prevent excessive sagging of the hoist line.

The skip tender waves his flag again, the skip loses headway, stops, swings pendulum-like for an instant, then starts the 700-foot descent to the canyon floor, dropping at the rate of about 300 feet per minute. Three minutes after leaving the Nevada rim of the canyon the skip has reached the canyon floor on the Arizona side. Had the men made the trip by highway, it would have taken from five to ten times longer. The signals of the skip tender for decelerating and stopping the skip are relayed to the hoistman by telephone by a signalman located in a lookout station on the edge of the canyon rim.

The cableways are of two types. One type has stationary towers and the other has movable, self-supporting towers. A track cable, or cables, runs from head tower to tail tower and, ordinarily, is fastened rigidly to an anchorage on one side of the canyon and, through a take-up device, to a similar anchorage on the opposite wall. At Black Canyon the anchorage for the stationary type is a structural-steel member concreted in a crosscut at the end of a tunnel 30 or more feet long. The towers act as anchorages in the self-supporting, mobile type—a counterweight, thrust rail, and the weight of the tower taking the track-cable pull.

The power units and hoists are situated near the head tower in the stationary type and within the head tower in the movable structures. Each such assembly consists primarily of electric motors, a group of hoist drums, compressed-air units for hoist-drum brakes, and a control board for the cableway operations. In every mobile tower there is also an electric motor and winch for shifting it along its track. When both head and tail towers are movable, the winches in both are operated simultaneously by the head-tower operator through the same electrical hook-up. Movement is effected by an endless cable that is wound on the winch drum and reeved between fixed sheaves at both ends of the trackway and sheaves at the tower.

A carriage runs upon the track cable and carries the fall blocks from which the skip, concrete bucket, or other load is suspended. The principal parts of the carriage are the steel frame, traveling wheels, sheaves for dump and hoist cables, and horn for carrying slack-rope carriers. Loads are conveyed along the track cable by the outhaul and inhaul cable, lowered and raised by the hoist cable, and dumped by the dump cable. The button line provides for spacing slack-rope carriers as required.

The outhaul and inhaul cable is a continuous line which starts from the tailward side of the track carriage, runs to and through two sheaves at the tail tower, returns above all cables to the outhaul sheave at the top of the head tower, and then passes downward to and around the conveying hoist drum, upward to and through the inhaul sheave, and outward to the track carriage. The hoist cable is reeved between the sheaves of the

†Assistant Engineer, U. S. Bureau of Reclamation, Boulder Canyon Project.

**Fourteenth of a series of articles on the Colorado River and the building of Hoover Dam.*

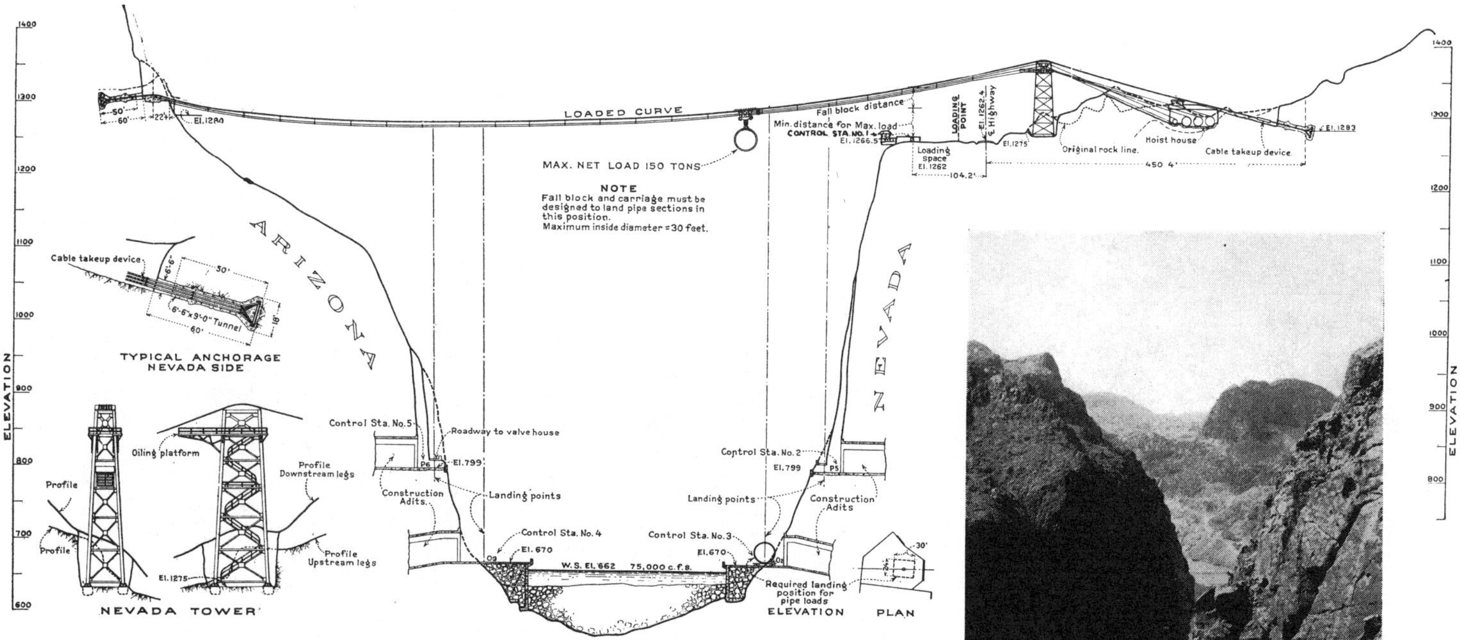


Courtesy, U. S. Bureau of Reclamation.

**Profile of the Government 150-ton cableway.**

Photo by B. D. Glaha

**Looking down from a cableway signal tower upon a skip in mid-air. Below is the lower cofferdam under construction.**

**Surveyors locating Nevada anchorage for a small cableway above the diversion-tunnel inlets.**

load fall block and of the carriage, runs to the hoist sheave in the head tower, and passes downward to the hoist drum. The dump cable leads from the drum of the dump hoist through the head-tower sheave to the track carriage, and thence to the dump block on the concrete bucket or other load.

The button line is fastened to the tail tower and runs just below the outhaul cable through a sheave at the head tower to a counterweight that holds the line taut. At intervals along the line are forged-steel buttons which engage or detach, one at a time, the slack-rope carriers that are carried by the track carriage. Thus, when the carriage travels away from the head tower, a fall-rope spacer is detached at each button, and, when the carriage returns, the spacers are picked up by the horn on the carriage. The carriers are provided to support the hoist and dump lines and to prevent them from sagging excessively between the head tower and the track carriage.

Among the first equipment placed in the canyon by Six Companies Inc. were several cableways, one above the outlet portals of the inner diversion tunnels and two of the same size across the upper portals. Later, after the Nevada section of the Black Canyon Highway had been extended to the top of the dam site, another service cableway was built above the sites of the canyon-wall valve houses—the head tower being located in a bay on the highway 800 feet from the top of the dam site. The hoist houses for these installations were placed in or near the head towers, or, in one instance, in the canyon near the river's edge.

The two cableways at the upper portals have been employed to move men and materials across the river, to aid in the construction of the foundations for the trash racks, to transport plate girders and steel members to the 50x50-foot steel gates, to assist in the erection of two double-track steel truss railroad bridges across the inlet portals of the Nevada diversion tunnels, and to carry workers to their positions on the canyon walls where they have been removing loose and projecting rocks. The cableway at the outlet portal has been used in the erection of the Stoney gates, while the one at the canyon rim has served to move trucks, power shovels, and other excavating machinery to the sites of the canyon-wall valve houses and the connecting outlet tunnels.

The dam, spillways, intake towers, and power house will occupy an area extending 2,000 feet along the canyon and 1,000 feet to each side of it, and will require more than 150,000,000 pounds of steel and 4,000,000 cubic yards of concrete for their construction. In order to provide ready access to all structures and to place these large amounts of materials in the shortest practical time, Six Companies Inc. has also built a group of five 20-ton cableways equipped with 3-inch diameter track cables and movable, self-supporting towers. These cableways are numbered

The steel trestle that supports one end of the trackway for the head tower of cableway 7.

Looking up the Nevada cliff at the head towers of cableways 7 and 8.

5 to 9 inclusive, from north to south. All head towers are on the Nevada side of the canyon.

Cableways 5 and 6 occupy the same runways on the landward sides of the spillway channels and will convey concrete for the construction of the two 150x650-foot spillway channels, the connecting 50-foot-diameter spillway inclined tunnel, the four 30-story, 75-foot-diameter intake towers, and the upstream portion of the main dam structure. Each has a span of 2,575 feet; and the maximum sag at the center when loaded is 151 feet—the distance from the lowest point of the cable to the crest of the dam being approximately 57 feet. The towers are of structural steel, 90 feet high, 46 feet in base width, and 32 feet in length along the track. They are parallel-traveling, and each is moved along its 600-foot runway by a Lidgerwood winch operated by a 100-hp. motor. The tracks on the Nevada side are almost entirely on rock, but those on the Arizona side are placed on a 130-foot fill made up of nearly 500,000 cubic yards of muck excavated from the spillway channel. They are of standard gauge, of 110-pound rails, and are spaced 45 feet 8 inches apart. Eight trucks, each having two pairs of 33-inch steel wheels, support each tower. The counterweight at the rear of every head tower contains 230 cubic yards of concrete weighing 929,300 pounds, and that at each tail tower contains 260 cubic yards of concrete weighing 1,052,000 pounds. The latter has a length of 48 feet and is 12x12½-feet in section.

The pull of the cable that tends to move the towers horizontally toward the canyon is counteracted by trucks, attached to the towers, which bear against a 100-pound rail placed with its web in a horizontal plane. The rail is fastened to an 8-inch timber which bears against a concrete curb. Steel rods, 2 inches in diameter and 13 feet long, hold the curb in position—the rods being sulphured in place in holes drilled in rock or connected to a continuous "deadman" of reinforced concrete in the fills. The horizontal trucks for each tower are 19 feet long and have ten 30½-inch double-flanged steel wheels.

The conveying and hoisting equipment consists of a 500-hp., 2,200-volt Westinghouse motor and of a 3-drum Lidgerwood hoist. The 1-inch-diameter inhaul and outhaul cable is wound on the rear drum, the ⅞-inch dump cable on the center one, and the ⅞-inch hoisting cable on the forward drum. Two 50-cubic-foot Ingersoll-Rand Type 30 two-stage compressors, operated by 10-hp. motors, are used for braking operations. The hoist drums are driven direct from the motor, which is reversible. The brakes on the hoist drums are of the external band type, and the friction clutches consist of wooden V blocks and steel friction cones. The band brakes are air released and counterweight set. Overload and overspeed devices set the brakes automatically.

The towers are in the shape of a right pyramid with the back leg vertical; are built of heavy structural steel; and are inclosed in corrugated sheet iron. The floors are of checkered steel plate; and, in order to eliminate the fire hazard, no wood is used. The towers and cableways are designed for normal operation with 20-ton loads and for maximum operation with 40-ton loads. The counterweights above the rear trucks of the movable towers have sufficient mass to balance the vertical pull of a 60-ton load on the cables. The connections between the trucks and the towers are effected through ball-and-socket and rocker joints to prevent the setting up in the towers of secondary stresses arising from inequalities in the cableway tracks or structural inequalities in the fabricated units.

Concrete for the downstream portion of the dam and the connecting structure between the power-house wings will be placed with the aid of duplicate cableway installations 7 and 8, which occupy the same runways at an average elevation of 113 feet above the crest of the dam. The head and tail towers are radial-traveling—the runway on the Nevada side being approximately 520 feet long and that on the Arizona side 450 feet.

The span between the towers is 1,405 feet, and the distance between the lowest point of the cables and the crest of the dam is 88 feet. Each head tower is 75 feet high, and each tail tower 42 feet. The runways are all in cuts, except where structural-steel trestles or concrete foundations have been built to cross ravines or to increase their lengths. On the Nevada side, old Lookout Point was removed in excavating the cut for the runway; and, in order to join the coverage of cableway 6, the runway was extended 100 feet northward by a steel trestle 56 feet high.

The horizontal pull on the head tower induced by the cableway load is resisted by super-elevating the front tracks on an approximately 25° incline from the horizontal. The tracks have the same gauge as those of cableways 5 and 6, and the distance between the center line of the tracks is 42 feet 3 inches. Other features of the equipment, dimensions, size of counterweights, and details of the cables are practically the same as those of No. 5 and No. 6, the power units and hoists being duplicate installations.

Cableway 9 is a radial-traveling type with the head tower fixed and the tail tower movable throughout a spanning radius of 1,374 feet and an arc of approximately 600 feet. This installation covers practically the entire area of the power-house wings. The runway track of the tail tower is 6 feet below the elevation of the crest of the dam, and the base of the head tower is 56 feet above the dam. The tail tower is 17 feet high, and its

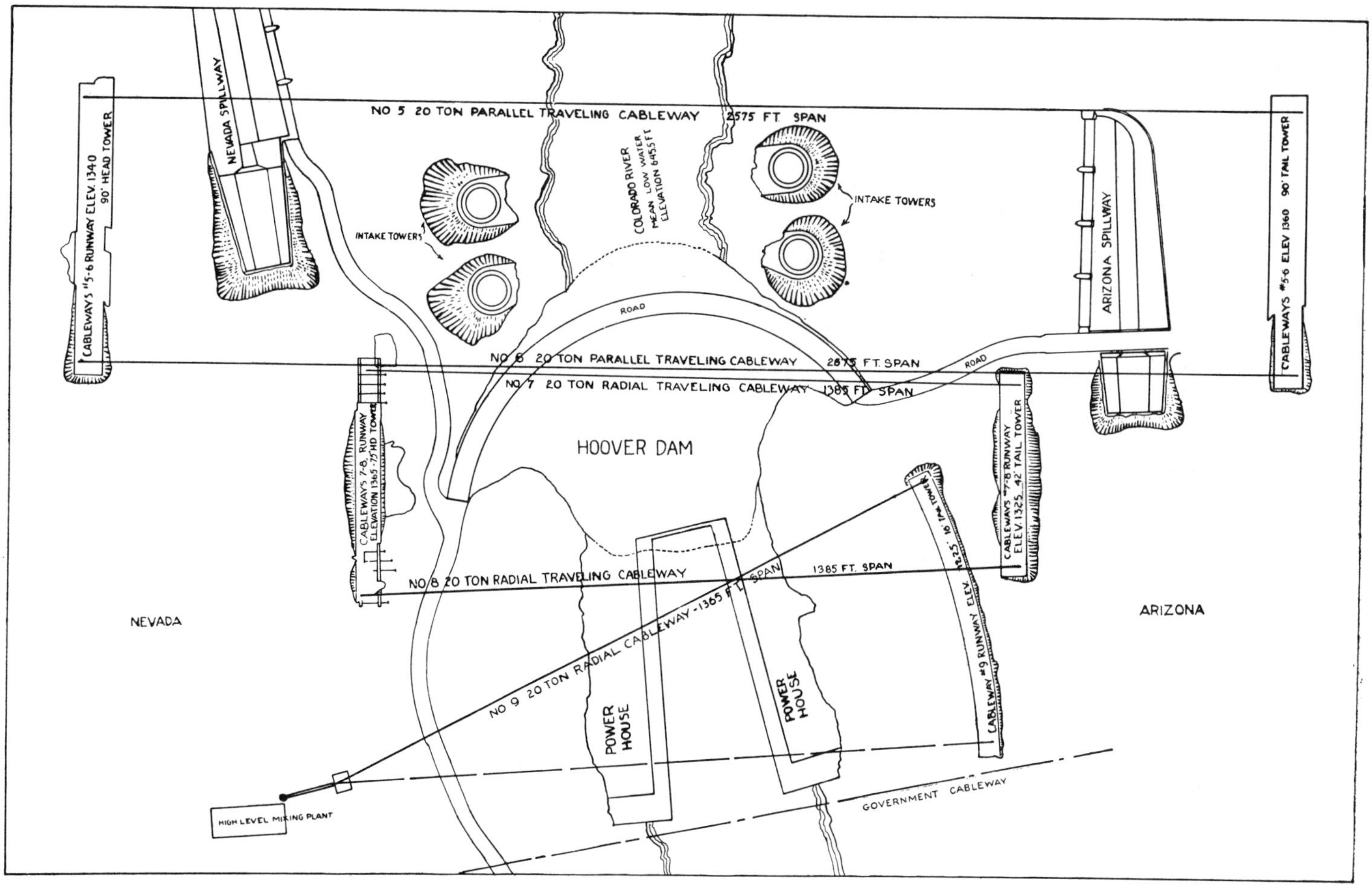


**Diagram showing the locations of the five 20-ton cableways which will play a leading part in all subsequent construction activities.**

**Head tower, carriage, and skip of a small cableway which crosses the canyon above the valve houses.**

counterweight contains 98.6 cubic yards of concrete weighing 400,000 pounds. The horizontal pull on the tower is resisted by a central horizontal rail in a manner similar to that in cableways 5 and 6.

The head tower is an A frame of structural steel 98 feet high and 32 feet in base width. The two side columns are connected by lattice bracing and supported through pin connections by steel shoes. The tower is braced near its top by a structural member which bears, through a hinge pin, against a base plate set in the solid rock of the cliff on the north side of the structure. The track cable and tower are guyed to the track anchorage by two 2¾-inch guys, the anchorage consisting of four 3½-inch steel rods and a structural-steel cross member concreted in a long tunnel and crosscut. The hoist house is located at the base and on the riverward side of the head tower. The power and hoisting equipment is of the same size and type as that of cableways 5 and 6. The travel of the tail tower is controlled electrically by the operator in the hoist house.

The concrete for Hoover Dam will be made at two locations: at the Lomix plant in Black Canyon, 4,000 feet upstream from the dam site, and at the Himix plant at the end of the United States Construction Railway on the Nevada rim and approximately at the elevation of the crest of the dam. Roadbeds have practically been completed for the railroads both from the Lomix plant to the dam site, running beneath cableways 5 to 8 inclusive, and from the Himix plant to the Nevada spillway, running beneath all cableways. At the time of pouring the main dam structure, the contractors plan to mount 8-cubic-yard bottom dump buckets on flat cars to transport the concrete from plants to cableways and by cableways from cars to pouring sites. Each bucket with its load will weigh about 20 tons.

The dam will be poured in vertical columns, approximately square in section and varying in side dimensions from 60 feet at the upstream face to 25 feet at the lower face. The concrete is to be placed in horizontal layers not exceeding 5 feet in thickness, and the tops of the columns will at no time differ more than 35 feet in elevation. The rate of placing the concrete is restricted by specifications to a pour of not more than 5 feet in depth in 72 hours and not more than 35 feet in depth in 30 days. On this basis, at least 21 months of continuous mixing will be required to pour the 730-foot dam, or an average for the 3,400,000 cubic yards of 162,000 cubic yards per month. In addition, nearly 700,000 cubic yards has to be placed in the appurtenant structures.

Taking into consideration other factors, and assuming average working conditions, the concrete plants and cableways will be required to operate with slight cessation from the summer of 1933, the time the first concrete will be poured in the main structure of

**From left to right are the Nevada towers of cableways 8, 7, 6, and 5. Railroad tracks will be laid on the ledge in the foreground.**

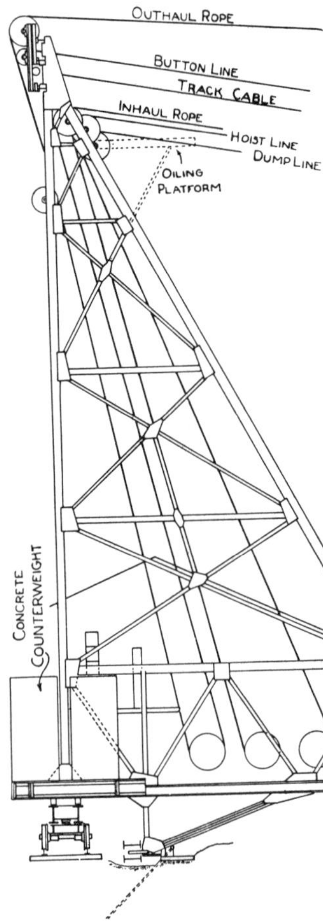


**Side elevation of a 90-foot head tower, showing arrangement of the various cables.**

the dam, until the project is completed in 1937.

The five cableways last described present unusual features of interest, but their capacity to move heavy loads is dwarfed by the huge Government installation which is to handle the steel penstocks and outlet pipes, the 30-foot-diameter steel penstock headers, and the turbines and machinery of the power plant. This cableway will be capable of conveying, lowering, and raising 150-ton loads. Its track will consist of six 3½-inch steel-wire cables spaced 18½ inches apart in a horizontal plane and supported on a structural-steel tower on the Nevada side and on a steel-and-concrete saddle on the Arizona side—the track cables being connected at each end to eyebars embedded in concrete anchorages.

The carriage will be operated at a speed of approximately 240 feet per minute, and will have means for acceleration and deceleration—the total length of travel being 1,050 feet of the 1,200-foot span. The load speed of hoisting and lowering will be about 30 feet per minute for 40- to 150-ton loads and 120 feet per minute for lighter loads. Provisions will be made for creeping speeds when conveying, and inching speeds when raising or lowering.

Cableway operations will be controlled from five stations. No. 1, for the chief operator, will be located in a lookout projecting 20 feet beyond the edge of the canyon wall and supported by steel I-beams anchored in a concrete foundation. The four other stations will be at the landing platforms at the portals of the construction adits which lead to the penstock header tunnels. The chief operator will direct the conveying of the load from the head tower out and over one of the other stations, when the control will be transferred to the designated station where an operator will direct its lowering or raising, having it in his power to move it at creeping speeds within a restricted zone. Transference of control between stations will be made with all equipment deënergized and all cable movements halted. Resumption of operations will be made by means of a definite interlock with the station assuming control. Each station will be provided with red and green signal lights to show which particular one is energized. All stations will be connected by a telephone system.

The hoisting machinery will be installed in a house located between the head tower and the anchorage on the Nevada side of Black Canyon. Each drum of the hoisting and conveying equipment will be run by a separate motor. Each hoist drum will be of sufficient diameter and length to wind the cable in a single layer. Heavy-duty and energizing brakes will be provided, and these will be capable of sustaining the full load under all conditions of operation, including interruption of the power supply. The lowering of heavy loads will be restrained by electric regenerative braking, and the loads are to be decelerated by dynamic braking before the application of the service brakes. These will function automatically in case of power-supply failure—decelerating promptly and holding the maximum load. Each drum will have an emergency brake that will automatically decelerate and hold the load in the event of overspeed or other contingency. The structural-steel head tower is 100 feet high and 25x35 feet at the base. The height was established by the allowance that would permit the largest load to clear with safety a parapet wall in front of the tower.

Operation and control will be by direct current generated by a 3-unit motor-generator set. The 300-hp., 1,000-revolution per minute motor will receive energy from a power station supplying 3-phase, 2,200-volt current. The generator will have a rating of 250 kw., 1,200 revolutions per minute, and 550 volts. The exciter will be of the 4-pole. 15-kw., compound-wound type.

Each of the three driving motors will be separately excited and shunt wound. The conveying motor will be rated at 400 hp. and 900 revolutions per minute, and each of the hoist motors at 175 hp. and 225/900 revolutions. Limit switches will govern the extent of travel of the track carriage and the fall blocks and will control the conveying, hoisting, and lowering operations.

The anchorages at both ends of the cableway will be of the mushroom type, consisting of 6½x9-foot tunnels leading into chambers of approximately 18x18x10 feet. The bulbs and at least 50 feet of the tunnels will be filled with concrete. The metalwork will consist

Left—Erecting the track carriages of a 20-ton cableway.

Below—Three-drum Lidgerwood hoist installation for cableway 9, showing the counterweights and air cylinders for braking.

One of five similar Type 30 air-compressor installations which supply air for the hoist brakes on the five 20-ton cableways.

of six bar-and-pin units connected to a 13x13-foot grillwork of heavy structural steel in the bulb of the excavation. The take-up device for the track cables will be made up of six 60-ton hydraulic jacks, one for each cable, mounted in toggle joints and installed between the Nevada anchorage and the hoist house.

The steel operating cables will include two $1\frac{3}{8}$-inch conveying lines, two $1\frac{3}{8}$-inch hoist cables, and two 1-inch button lines. The conveying drum in the hoist house will be 8 feet in diameter and 10 feet $5\frac{3}{4}$ inches long, and each of the two hoist drums will be 13 feet in diameter and 17 feet long. The rotations of the conveying and hoist drums will be electrically equalized by two Selsyn generators and a Selsyn differential motor so that the load when being conveyed will remain the same distance below the track cables.

The track carriage will carry 48 rollers—eight for each track cable, all of 24-inch tread diameter. The diameter of the hoist sheaves on the carriage and of the two fall blocks will be 35 inches, the same as that of the sheaves on the tower. The two connected fall blocks will carry the load by a single pin and links—hooks for the purpose not being permitted. The button lines will have button projections at 100-foot intervals. Two sets of slack-rope carriers are to be provided, one set for each hoisting and inhaul cable.

The hoists will be equipped with heavy-duty service and emergency brakes. Service brakes will be of the solenoid-operated, spring-set type installed on the commutator side of the driving motor. Each hoist is to be equipped with an emergency brake having a 10-inch band lined with a $\frac{1}{2}$-inch thickness of Ferodo lining. This latter brake will be set by a heavy spring and released by an electrically operated thrustor that will enable the operator at any control station to set the brake at will as well as to cause it to be applied automatically in case of current failure. An overspeed device on each hoist will set the service and emergency brakes in the event of overspeeding or current failure. Owing to an adjustable speed device, the brakes cannot be applied other than smoothly and gradually.

When the cableway carries a load of 150 tons, the pull will be 2,058,000 pounds on the Arizona anchorage and 2,100,000 pounds on the Nevada anchorage. The components of this force at the head tower will amount to 880,000 pounds vertically and 29,000 pounds horizontally. The maximum tension on the inhaul and outhaul cable will be 47,000 pounds. Track cables have been designed with a factor of safety above 3, and conveying and hoisting ropes with a factor of safety greater than $4\frac{1}{2}$. The breaking load of each track cable is 1,070,000 pounds or 6,420,000 pounds for the six cables, and the breaking strength of the conveying as well as of the hoisting cable is 106,000 pounds.

The engineering department of Six Companies Inc., under the direction of General Superintendent Frank T. Crowe, Chief Engineer A. H. Ayers, and Office Engineer J. P. Yates, designed the movable towers and runway structures. The Lidgerwood Manufacturing Company, of Elizabeth, N. J., furnished the hoists and cableways for the five 20-ton installations, and the Consolidated Steel Company, of Los Angeles, Calif., erected the movable towers. The contract for furnishing and erecting the 150-ton Government cableway was awarded to the Lidgerwood Manufacturing Company on its low bid of $172,110. Westinghouse motors are used in the Six Companies Inc. installations, and General Electric motors, generators, and electric equipment will be used in the Government cableway.

As has been mentioned in previous articles of this series, Hoover Dam is being built by the United States Government through the Bureau of Reclamation. Dr. Elwood Mead is commissioner and administrative head of the Bureau; R. F. Walter is chief engineer in charge of design and construction; Walker R. Young, as construction engineer, represents the chief engineer on this project and has direct supervision of the building of the dam and appurtenant works. Principal contractors carrying on the construction are Six Companies Inc. of San Francisco, and The Babcock & Wilcox Company of New York. The contracts awarded these two concerns are the largest ever let by the Bureau of Reclamation, and total approximately $59,-000,000.

# Construction of the Hoover Dam*

## A Description of the Tunnels for the Penstock Headers, the Penstocks, and the Canyon-Wall Outlets

RUSSELL C. FLEMING

**DOWN SHE GOES!**

The previous article described the aerial transport system that spans the canyon to facilitate the movement of men and materials. Here we see a complete Bucyrus-Erie 43-B power shovel being lowered into the abyss by means of two of the 20-ton cableways working in unison.

OPERATIONS at the Hoover Dam have reached an important transition period. Until the spring of this year efforts had been directed mainly towards works in connection with the diversion of the river or otherwise preliminary to the construction of the dam, itself. Now most of those preliminaries have been executed and activities are centered on the permanent features of the project—those that are necessary for the generation of power and for the control of the Colorado during flood periods, those at which tourists will marvel and which they will photograph for many years to come. The transition is not sharp. It cannot be said that the preliminary operations were finished one month and construction on the dam and the other permanent features begun the next, for much of the work is necessarily interrelated. Too, some of the work already completed will be converted later and become an integral part of the final structure. The first phase is just merging gradually into the second and more enduring phase.

During the late spring of 1933 there was a change in the scene of activities, and the contractors then took up the work in which they are now principally engaged and which includes the following: excavation of the river bed for the dam foundation; final trimming of the canyon walls where the dam abutments will rest; pouring concrete for the spillways: excavation for the intake towers; driving the tunnels for the penstock headers, the penstocks, and the outlet works; finishing the 150-ton cableway which will be permanent and which will be used during the building period primarily to convey the heavy power-house machinery and the pipe sections for the penstocks from the top of the canyon to the river bottom; and finishing the construction of as well as equipping the plant of the Babcock & Wilcox Company in which will be fabricated the enormous sections of pipe for the penstock headers, the penstocks, and the outlets.

A comprehensive view of the whole project is necessary in order to understand the part that these works will play after the dam is completed. A brief sketch, previously published in this series of booklets, is therefore given again so that it will be easier to follow the operations now in progress. When the dam is finished and the reservoir filled, some provision must be made to allow excess water to escape to the river channel below. Some of the water will course through the turbines—expending its force in generating electric power and then discharging into the tailrace. Means are also being provided for water to by-pass the turbines and to flow into the stream bed. In addition to this, however, spillways are very necessary at the top of the dam to permit the escape of the excess of inflow over outflow. In the usual dam construction the spillways are built in the top of the dam or in the wings, and an open channel, carefully contoured, carries the water which overflows the spillway weirs to the river channel below.

**Fifteenth of a series of articles on the Colorado River and the building of Hoover Dam.*

DRILLING THE SPILLWAYS

The digging of the spillways on either side of the river involves the removal of 580,000 cubic yards of rock. For placing the accessory structures there will be required 110,000 cubic yards of concrete. Each spillway will connect, through an inclined tunnel, with the outer 50-foot diversion tunnel on its side of the river, thereby making possible the by-passing of 400,000 cubic feet of water a second should this ever become necessary as the result of a flood of sufficient volume. At capacity, the discharge from each spillway will be equal to the normal flow over Niagara Falls, and the total drop will be more than three times as great. The energy of the falling water will be about 25,000,000 hp. The picture at the upper right shows "Jack-hamer" men starting the excavation for the Nevada spillway. Below is a view of the upstream portion of the weir foundation in this spillway as it appeared several months later. Above this legend is the excavation for the Arizona spillway as it looked on April 21, 1933.

In Black Canyon a slightly different plan is being followed. Two spillways are being constructed, one on each side of the dam. The weirs and weir channels are set high up on the rocky walls of the canyon with the crest of the weirs paralleling the course of the river. Large excavations were made in the solid rock of the walls for these spillways. The concrete-lined open channel behind each crest will be approximately 650 feet long, 150 feet wide, and 120 feet deep, and from these channels the water of the two spillways will be discharged through the two outer diversion tunnels. Water from the Nevada spillway will discharge down an inclined tunnel into the outer (No. 1) diversion tunnel on the Nevada side and the spillway on the Arizona side will discharge similarly into the outer (No. 4) diversion tunnel on that side. Thus portions of each of two of the diversion tunnels will become permanent spillway channels when the time comes.

It is estimated that 110,000 cubic yards of concrete will be used in the spillway works—28,000 in the channel linings and the rest in the weirs, and a total of 580,000 cubic yards of rock has been excavated for the two spillways. The weirs on each side will be about 450 feet long, with an effective length of 400 feet. Each weir will be topped by four 100x16-foot drum gates. Behind the weirs the open channel will slope downstream on a 12 per cent grade to an overflow weir, 35 feet high, and then connect with the inclined tunnel leading down to the diversion tunnel. The inclined tunnel will be 50 feet in diameter (the same as the diversion tunnel) where the two tunnels join and, as the incline rises, will expand first to 60 feet and then into a section with a 3-center top arch of 80-foot span. The lower part of the incline is laid on a parabolic curve, and involves precise office and field engineering.

On both sides of the canyon the work of excavating for the weirs and open channel was completed by the first of May, and forms were being placed for the concrete lining. On the Nevada side the inclined tunnel was excavated but not yet lined with concrete, and on the Arizona side a 7x14-foot pilot raise had been driven through to connection with the diversion tunnel. When the outer diversion tunnels are no longer needed for diversion purposes, each will be stoppered with a solid concrete plug about halfway down its length and just above where it connects with the spillway tunnel. The diversion tunnels are enlarged at that point and the plugs, 400 feet long, will be firmly anchored in the solid rock. The course of the outer diversion tunnels was purposely laid so that those tunnels could be made a part of the spillway structures.

Of prime importance in the completed project will be the works that are designed to lead the water from the reservoir to the turbines in the power house. These will consist of four intake towers in the reservoir above the dam, of a penstock header leading from the bottom of each tower past the dam, and of four penstocks branching from each header. Each of the latter penstocks, excepting one, leads to a turbine in the power house. Also included in these plans are outlet works below the penstocks whereby the water may be made to by-pass the turbines and to discharge directly into the tailrace. At the present time the sites for the intake towers have been excavated and the tunnels for the upper penstocks, the penstock headers, and the outlet works are being driven. These tunnels will contain steel pipes fabricated from plate steel in the plant of the Babcock & Wilcox Company located 1½ miles from the dam site.

Description of the intake towers must be left to a later article, likewise the absorbing story of how the great pipes will be fabricated, transported, and placed. Of more immediate interest is the driving of the tunnels for this piping. As the power house is to be U-shaped, with one wing on each side of the river, and as each wing will contain half of the power-generating equipment, the penstock works will also be twinned and almost identical in each

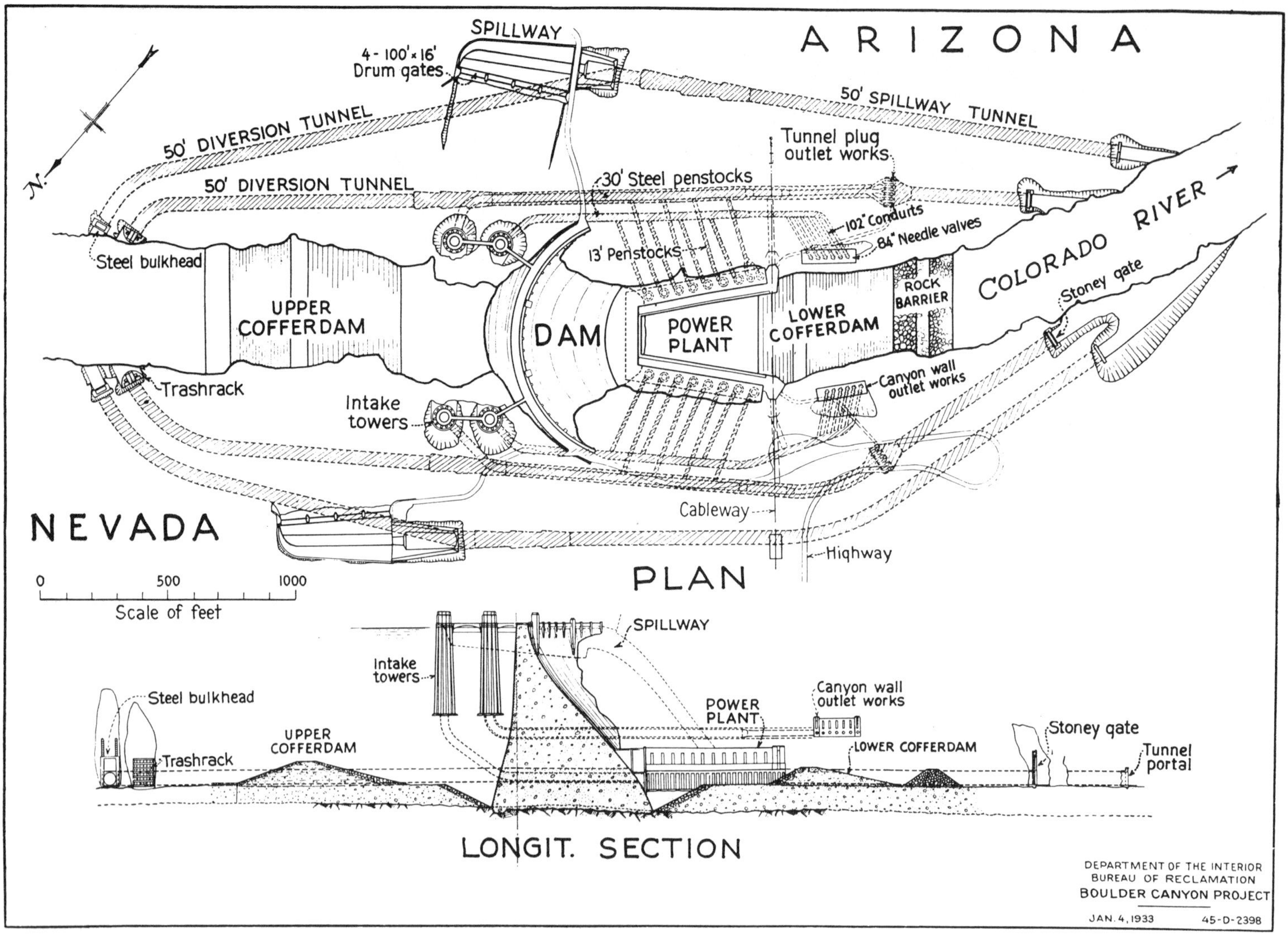


GENERAL SCHEME OF OPERATION

These diagrams show the relationships between the various structures. Reference to them and to the drawing at the bottom of page 20 will enable the reader to follow the text understandingly.

canyon wall. A description of the layout on one side will therefore suffice for both.

A pair of similar intake towers will be built on each side. From each upstream intake tower is being excavated an inclined tunnel, 41 feet in diameter, to connect that tower with the inner diversion tunnel (No. 2 on the Nevada side and No. 3 on the Arizona side). Dropping directly from the bottom of the tower, this tunnel bends gently to form an incline and then, at the lower end, bends again to meet the diversion tunnel. The incline will be lined with 2 feet of concrete, giving a finished diameter of 37 feet, and through it and the inner diversion tunnel will be run the 30-foot steel pipe line connecting the tower with the penstocks and the outlet works. The diversion tunnel will be sealed just above the connecting point with a concrete plug 300 feet long. Provision was made for this plug in driving the diversion tunnel—the concrete lining, except for the arch section which was left unlined, being poured in conformity with the shape of the plug. In the meantime, wooden forms have been placed in the invert and side-wall sections so as to provide a continuation of the smooth inner face of the tunnel. These will be torn out when the plug is to be put in position. The plug will be approximately 65 feet in diameter at the widest point and will have three tapered shoulders whereby to anchor it to the tunnel wall. Thus the two inner diversion tunnels will also find employment in the ultimate structure.

From each of the two downstream intake towers a header tunnel is being driven all the way to the penstock connections and beyond to the canyon-wall outlet works. This tunnel also is to be 41 feet in excavated diameter and to have an inside diameter, including the concrete lining, of 37 feet. To drive the tunnel and to admit the pipe sections it was necessary to run a construction adit 26 feet wide and 43 feet high from the canyon below the penstock tunnels, at elevation 820, to the line of the tunnel. On the Nevada side the header will be 1,322 feet long from the construction adit to the intake tower and require the removal of approximately 75,000 cubic yards of rock: on the Arizona side the tunnel will be 1,197 feet from the adit to the tower and require about 67,000 cubic yards of excavation.

Each of these 41-foot tunnels is being driven full face without the aid of a pilot bore by means of a huge drill carriage or jumbo which is mounted on wheels and which is moved on rails. This device, which makes it possible normally to combine as many as 20 drills in a concentrated attack upon the breast, is a modification of the type of carriage that was developed for driving the diversion tunnels and that was described in detail in an earlier article in this series. The carriage has three working platforms in addition to the ground level. Across its front are four horizontal bars, on each of which are mounted from four to six Ingersoll-Rand N-75 drifter drills. Compressed air is furnished by the established plant and delivery system and is fed through a 4-inch line to the piping on the carriage. A standard round of drilling, as illustrated in an accompanying sketch on page 20, is used at the tunnel face. The firing is done with delay primers. The normal rate of advance is one 12-foot round per shift. The rock is handled at the face by electric shovels and taken out of the construction adit in trucks. It is dumped on the canyon floor, picked up by another shovel, and hauled away. Part of the rock is used in the cofferdams and for construction purposes in the river bed. By May 1, the

## WATERWAYS THROUGH ROCK

For a portion of their lengths the diversion tunnels next to and on either side of the river will serve as penstock headers, and from them the water will flow through four penstocks to the power turbines. The picture at the right shows one of these penstock tunnels being opened from within the lined interior of a diversion bore. A Conway mucking machine is at work.

Below is a view of the inclined spillway tunnel on the Nevada side while driving was in progress. This bore will carry excess water from the Nevada spillway to the No. 1 diversion tunnel. It was excavated by driving a top heading upward at an angle from the diversion tunnel. This pioneer opening was then enlarged until it was 14 feet wide and 56 feet high; and, as a final operation, the remaining material on either side was taken out to the full 56-foot circular section. This last stage, which is pictured, was carried on from the top downward, the muck being allowed to fall into the diversion tunnel from which it was removed for disposal.

## SHARPENING STEEL

Hundreds of tons of drill steel are being consumed at Hoover Dam. The reconditioning of drill steels is an important item in maintaining drilling progress. At the left is one of the many outdoor shops that have been set up for this purpose.

header on the Nevada side had been driven 695 feet and on the Arizona side 652 feet—each more than halfway.

These tunnels have no grade and are at elevation 820, considerably higher than the headers in the diversion tunnels. In the diversion tunnels the center line of the steel pipe will be approximately at elevation 649 where the penstocks take off, and the center line of the turbine runners in the power house will be about at elevation 637. Consequently, the penstocks from the upper headers will incline at a fairly sharp angle while the penstocks from the headers in the diversion tunnels will be almost flat.

Four penstocks will branch from each header, or eight from the two headers on each side of the river. These penstocks will reach the power house alternately—that is, Nos. 2, 4, 6, and 8 on the Nevada side will branch from the upper header and Nos. 1, 3, 5, and 7 will branch from the diversion-tunnel header. On the Arizona side the arrangement differs in that Nos. 1, 4, 6, and 8 will come from the upper header and Nos. 2, 3, 5, and 7 will extend from the diversion-tunnel header. The penstock tunnels are being excavated 21 feet in diameter and will be 18 feet in diameter after the concrete lining is in place. Each penstock pipe will be 13 feet in diameter. The Nevada penstock tunnels from the upper header will have an aggregate length of 1,373 feet and will require the excavating of 17,600 cubic yards of material, and the four from the diversion-tunnel header will be 1,473 feet long and 18,900 cubic yards of rock will be removed from them. The Arizona tunnels from the upper header will have a total length of 1,348 feet and involve the excavating of about 17,300 cubic yards, and those from the diversion-tunnel header will be 1,356 feet long and will call for proportionately more excavating.

DRILL CARRIAGES

Tunnel driving at Hoover Dam has been notable for the development of huge structures which permit a number of drills to work simultaneously on a rock face. These carriages not only provide for concentrated drilling but save time in setting up and also facilitate moving the drills back from the face before blasting and returning them after mucking is completed. The accompanying views show two such carriages or jumbos used in connection with the penstock work. Below is a rear view of the rail-mounted structure employed in the 41-foot headers. The drills are mounted on four horizontal bars across the front, permitting as many as twenty machines to operate at once and making it possible to drill the full face (about 130 holes) without moving the carriage. At the left is a truck-mounted carriage of the type used in the flat tunnels leading from the diversion bores to the turbine sites.

All the penstock tunnels on the Arizona side and those from the upper header on the Nevada side are being driven inward from the canyon. Those from the diversion tunnel on the Nevada side were advanced from within the diversion tunnel before it was utilized as a watercourse. The flat tunnels are being driven by truck-mounted drill carriages equipped with fourteen N-75 drifters mounted on crossbars and operated from the ground and from three platforms. The normal rate of advance is one 12-foot round per shift. A round usually consists of 35 holes, and these are distributed over the face much as are the holes in the penstock-header round, which is illustrated. Eleven boxes of powder are ordinarily used per round, and 10 cubic feet of rock is excavated per foot of advance. A tunnel crew includes four miners, four chuck tenders, one nipper, one operator for the Conway mucking machine, one cable tender, two pitmen, and two truck drivers per shift.

The inclines to the upper headings are excavated by driving an 8x8-foot center heading and then enlarging this opening to full size by two ring drillings. The usual round in the center heading consists of eighteen holes, and the normal advance per round is 6 feet. The ring holes are drilled on 3-foot centers. The center headings are kept just ahead of the full face so that the same set-up will do for both. The muck is pulled down the incline with a Bagley scraper and loaded with a Conway mucking machine into trucks for disposal. The mucking machines are also used in the flat tunnels—one mucking machine serving two tunnels.

At the last report all the Arizona tunnels were in process of driving except No. 4, which was completed. On the same side the penstock tunnels to the lower header will be driven close to the diversion tunnel, only an adequate rock barrier will be left standing between

TUNNEL MUCKING

The 41-foot circular penstock headers on either side of the river were drilled full face with jumbos and the muck was loaded into trucks by electrically operated shovels. The picture at the right shows a Marion 450 shovel at work.

them in each case. These several barriers will be holed through and connections made later when water no longer flows through the diversion tunnel. The Nevada tunnels to the lower header were driven full length from within the diversion tunnel before water was turned into it, and were finished late last year. Arched, reinforced-concrete bulkheads about 16 inches thick and with a 20-foot radius of curvature were put in place to seal the openings before the diversion tunnel was used as a watercourse. Each of the penstocks will bring water to one 115,000-hp. vertical turbine in the power house except No. 8, downstream on the Arizona side. This penstock will split and will serve two units of 55,000 hp. each—the tunnel also being split into two smaller ones near the outlet for this purpose.

The upper-header tunnels continue downstream from the penstocks and terminate in the canyon-wall outlet works. Below the penstocks the steel headers in diversion tunnels Nos. 2 and 3 also continue to a series of outlets in a tunnel plug about 650 feet downstream from the last penstock. All these outlets will permit water to by-pass the turbines and to discharge into the river channel below. The canyon-wall outlet works are at elevation 820, about 180 feet above the normal tailrace water surface; and, when water is passing through them, a solid stream 84 inches in diameter will issue straight out from each and fall that distance in a wide trajectory to the tailrace. In most of the sketches of the project, as it will appear when completed, these streams, issuing from far up on the canyon wall on each side, are shown as a conspicuous feature.

SCALING WALLS

Ever since Six Companies Inc. moved into Black Canyon the scaling of the walls of the gorge has been in progress. It has been necessary not only to remove loose and projecting bowlders but also, in many cases, to cut back into the solid rock itself to provide footholds, anchorages, and keyways of one sort or another. The scalers, inured to the hazards of their work, sit in bos'n's chairs suspended on ropes from high above. Here are shown a group of these workers scaling down the Nevada side of the canyon wall in the area where a part of the power house will be built.

The outlet works on each side of the river are substantially alike. Downstream from the construction adit the upper header tunnel is excavated to a diameter of 35 feet and, at the lower end, six horseshoe-shaped tunnels, 11x11 feet in section, split off and come out to the canyon wall. Beyond the penstocks the steel lining contracts to a diameter of 25 feet and terminates in the outlet tunnels in six pipes, each 102 inches in diameter. The outlet tunnels will not be lined until the pipes are in place, and then concrete will be forced solidly between the piping and the tunnel walls with air under pressure. Each outlet will be closed with an 84-inch internal differential needle valve. The tunnels for these works are nearing completion.

The steel headers in the diversion tunnels also contract to a diameter of 25 feet after they pass the penstocks and, several hundred feet lower down, end in a series of outlets in plugs in the diversion tunnels. At the plugs each 25-foot header branches into three 13-foot pipes and each 13-foot pipe branches into two 86-inch pipes. These six outlets pass through the plug in horizontal rows 16 feet apart. The six 72-inch hydraulic needle valves closing the outlets will be operated from a chamber above them. Access to the valve chambers will be had through horseshoe-shaped adits leading to the landing platforms of the 150-ton cableway in the canyon. Water through these outlets will flow down the diversion tunnel to the river. Through the turbines and these outlets combined will be discharged 120,000 cubic feet of water per second. Of this, 34,000 will be for maximum power generation and the rest will be valve discharge. By contrast, the maximum capacity of each spillway will be 200,000 cubic feet per second.

While these works are being carried to completion, the river bed is being rapidly excavated down to solid rock preparatory to the building of the dam, itself.

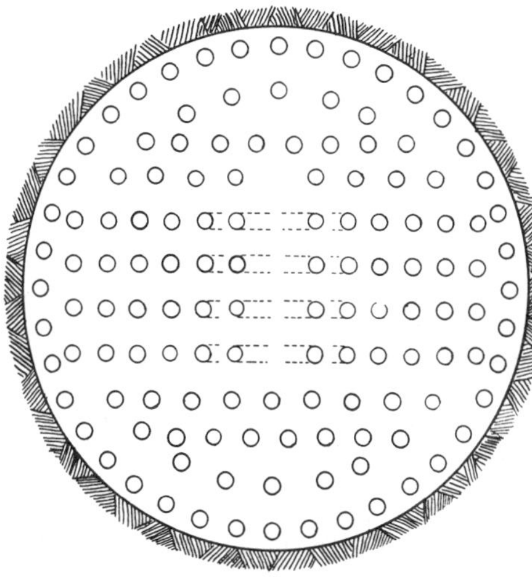

DRILLING ROUND

This diagram shows the approximate number and the positions of the drill holes in a normal round in the face of the 41-foot penstock header tunnels. The four inner cut holes in each vertical row on either side of the center are drilled at angles so that the opposing holes tend to converge at a depth of about 14 feet. The two outer and adjoining rows of cut holes are drilled at angles to depths of about 18 feet. After the inner and outer cut holes have been fired, the vertical rows on either side of them are shot, usually with four delays, thus breaking out a horizontal section the full width of the tunnel. Next are fired some of the horizontal rows above the center, followed by those below the center, the shooting terminating with the remaining holes at the top. After they are lined with concrete, these tunnels will be 37 feet in diameter.

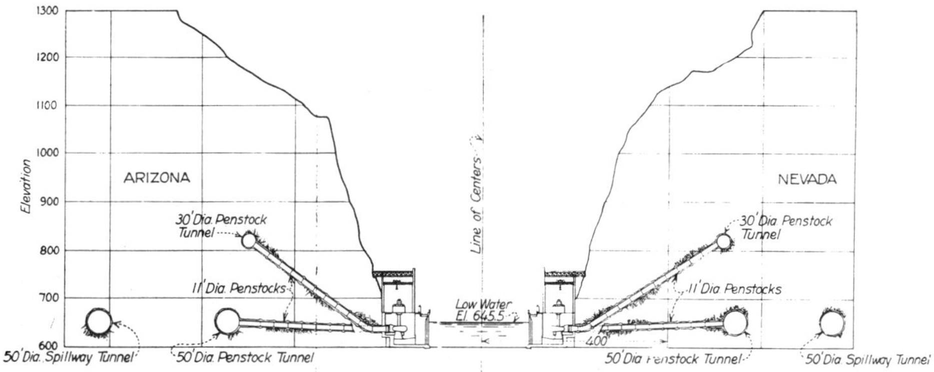


SECTION THROUGH POWER HOUSE

This Bureau of Reclamation drawing shows clearly the general arrangement of the tunnels and power houses. The dam footings will be approximately 140 feet below the base level shown here, and the crest of the structure will be at elevation 1,282.

# Construction of the Hoover Dam*

## How the Concrete Is Being Cooled as It Is Poured

LAWRENCE P. SOWLES†

THE ROOTS OF HOOVER DAM

General upstream view of the dam and Black Canyon as they appeared from the Government cableway on September 15, 1933. The division of the dam into a large number of columns for the pouring of the concrete by stages is clearly shown. The tops of the highest blocks are at elevation 640, which was about 6 feet below the mean low-water surface of the river before it was diverted. The top of the dam will be at elevation 1232, or 592 feet higher than shown here. At the center, in the foreground, is the inclined trestle on which the cooling-water mains enter the structure at elevation 560. This leads to the 8-foot slot, in the center of the dam, which can be easily traced. Close inspection of the right central area will disclose the fall lines from several of the overhead spanning cableways. At the left is the railroad, at elevation 720, over which some of the concrete is delivered from the Lomix plant just beyond the tunnel. A suspension bridge by which workmen cross the gorge shows at the top. The shovel on the upper cofferdam serves as a head tower for a "Joe McGee" bridle cableway which delivers materials on to the dam.

THE plans and specifications prepared by the U. S. Bureau of Reclamation for the guidance of the contractors provide that the concrete in Hoover Dam shall be progressively cooled as it is poured and that a refrigerating plant shall be maintained for that purpose. It is usual to associate power shovels, rock drills, trucks, locomotives, and countless other mechanical agencies with the construction of a large dam, but the employment of refrigeration is, indeed, novel.

In view of the fact that some fairly large concrete dams have been built without the aid of artificial cooling and are today successfully holding back great volumes of water, the question naturally arises: Why was it considered necessary to include this unusual specification in this case, even though Hoover Dam is of unprecedented size? It will be the purpose of this article to answer this question, and also to describe the plant which the contractors have set up to do the prescribed work.

We are all familiar with the fact that bodies expand when they gain heat and contract when they lose heat. We have seen the bitumastic filling squeezed out of concrete pavement joints in the summer and have noticed those same joints open to a considerable distance in the winter. Similar changes take place in all structures which undergo temperature changes. Sometimes they are not of sufficient magnitude to be visible to the eye, but they are just as surely present. The pavement referred to is given room in which to move and thereby to relieve the tension or compression. If, however, there is no provision for movement, the effect of temperature changes is to set up internal stresses. The magnitude of these stresses varies according to the range of the temperature changes and the capacity of the body to adjust itself within so as to accommodate the internal strains. When the temperature stresses, plus the other stresses to which the body is subjected, exceed the resisting capacity of the structure, they manifest themselves in the form of cracks. These fractures are essential outbursts of strains, which must find relief either internally or externally. This cracking may or may not be of serious consequence, depending upon the service of the structure affected. In extreme cases it amounts to failure.

A dam of the gravity type, such as Hoover Dam will be, resists the pressure of the water behind it because its large mass is placed so as to give it the necessary stability. Such a huge volume of concrete undergoes seasonal and other temperature changes which set up enormous internal stresses. The greatest of these changes occurs during and shortly after construction, and arises from the setting action of the cement. Portland cement is a complex mixture of oxides, silicates, and carbonates of various elements. When water is added to it there is a chemical change, called hydration, during which are formed

†**Designer, Engineering Dept., Six Companies Inc.**

***Sixteenth of a series of articles on the Colorado River and the building of the Hoover Dam.***

CONCRETE PATCHWORK QUILT

Two pictures made from the upper cofferdam, the lower a telephoto. The black structure in the background is the cooling tower over which 6,000 gpm. of water can pass. To its right and nearer the eye is the refrigeration plant. Horizontal keyways for knitting the dam together show on some of the blocks of concrete. The derrick boom at the base of the upper picture is 180 feet long and will be used for pouring concrete for the Nevada intake towers.

crystals which interlock among themselves and with the aggregates of the concrete to form a unified mass with high compressive and shearing strength. This process of hydration, which develops the strength of the concrete, serves to liberate a considerable quantity of heat. If this heat is not removed it raises the temperature of the concrete mass.

If the dimensions of the dam are reasonably small, this heat is soon dissipated by radiation and exerts no injurious effect upon the structure. Sometimes, to prevent the internal stresses from forming cracks upon the surfaces, sufficient reinforcing steel is added to give the structure the necessary resistive strength. Thus, while the temperature effects are present, they are absorbed internally. In large, unreinforced-concrete dams the heat of setting passes out by radiation from the surfaces exposed to the air and by conduction into the rock abutments, and since such unreinforced concrete is incapable of resisting tensile stresses, the shrinkage resulting from the cooling produces surface cracks. Except in the case of dams of the very smallest proportions, means are usually provided for filling these cracks at periodic intervals with grout, a neat mixture of Portland cement and water. Since the cooling progresses slowly, grouting may take place considerably after the structure is completed, the interval depending upon the shape and dimensions of the mass and the temperatures of the surrounding bodies to which it transfers its heat. If grouting is done before the cooling is complete, subsequent cooling of the mass towards its point of thermal equilibrium may produce further cracks which cannot then be grouted, and these fractures may endanger the safety of the dam. In any event, the cracks are usually unsightly and uneconomical.

In a structure as massive as Hoover Dam, the heat must travel so far to reach the surface that an inordinate time would be required for cooling to progress to the point where grouting could be resorted to to fill cracks and thereby stop any existing or impending leakage. Indeed, in this instance, it has been computed that 125 years would have to elapse before grouting could be done with complete assurance that there would be no further cracking. To obviate this long wait and to exercise the greatest possible precaution towards making the structure immediately and permanently safe, it was decided: (1) to facilitate the dissipation of the setting heat by dividing the dam into a number of independent blocks and pouring these in definite stages in columns keyed together to form one compact mass when completed; and (2) to cool the entire structure by refrigeration to a point where there will be no subsequent temperature drop of sufficient magnitude to produce crack-forming strains. By following this procedure the cooling will be effected quickly, and upon its completion grouting can be done with every assurance that the individual blocks will be unified into a strong, secure, and enduring monolithic structure.

To determine the amount of cooling that would be required to reduce the temperature of the concrete to a safe point, the Bureau of Reclamation conducted extensive researches into the thermal properties of concrete. It was found that the average temperature rise resulting from the setting reaction would be approximately 40°F., and that the quantity of heat to be removed would be approximately 965 Btu's per cubic yard of concrete for each degree that it was to be cooled. This latter figure will vary slightly depending upon the

TRANSPORTING CONCRETE

Telephoto (left) of the Nevada footing and of the railroad at elevation 720 while two of the cableways are lifting buckets of concrete from cars. Note how the railroad is shored up. When built, the line rested on a fill at the edge of the river. The refrigeration plant is at the left, in line with the railroad.

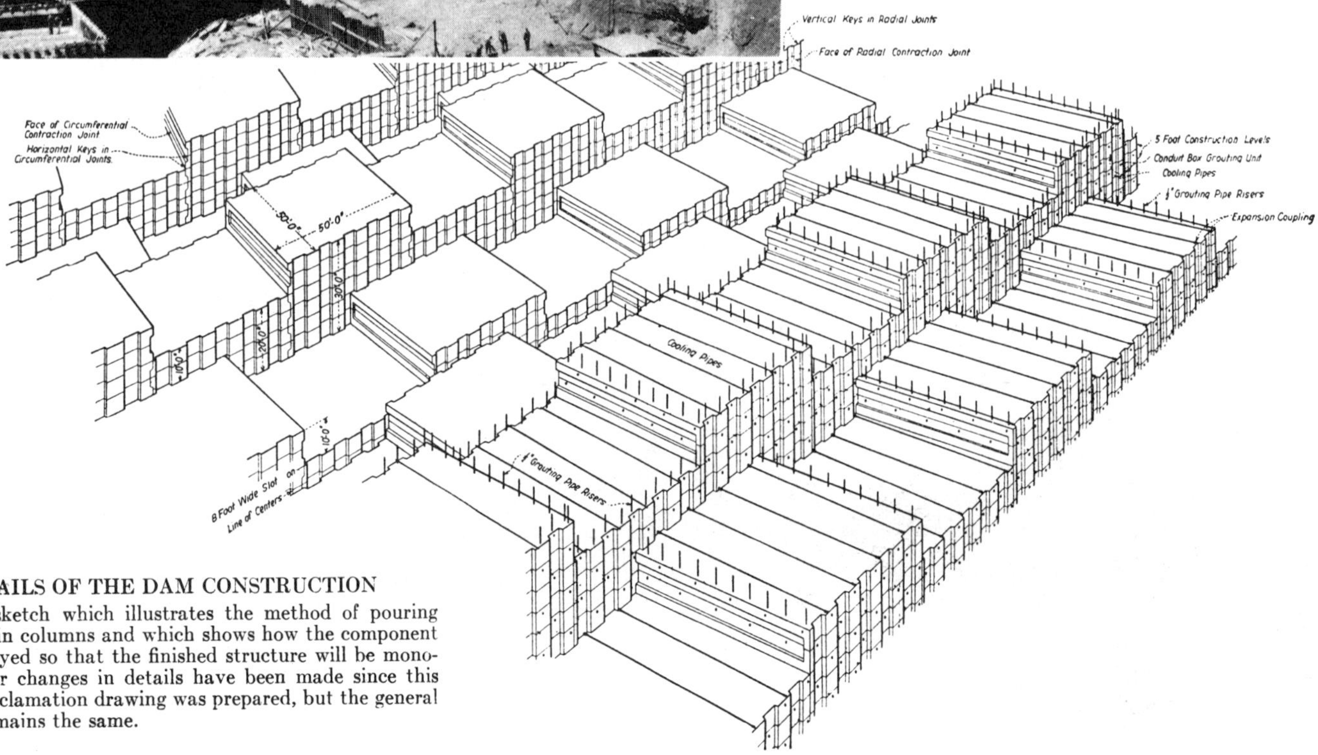


DETAILS OF THE DAM CONSTRUCTION

Perspective sketch which illustrates the method of pouring the concrete in columns and which shows how the component blocks are keyed so that the finished structure will be monolithic. Minor changes in details have been made since this Bureau of Reclamation drawing was prepared, but the general procedure remains the same.

PUMPS AND AMMONIA COMPRESSORS

**Each of the three compressors (opposite page) has a capacity of 275 tons of refrigeration. They were used as air compressors during the early construction stages and then converted for their present service by substituting ammonia cylinders. Note the all-welded piping, which was photographed before it was insulated. The pumps in the foreground of the picture at the right handle precooling water. Beyond them are the refrigerated-water pumps. The horizontal cylinders in the background are the coolers in which heat is extracted from the refrigerated water returning from the dam.**

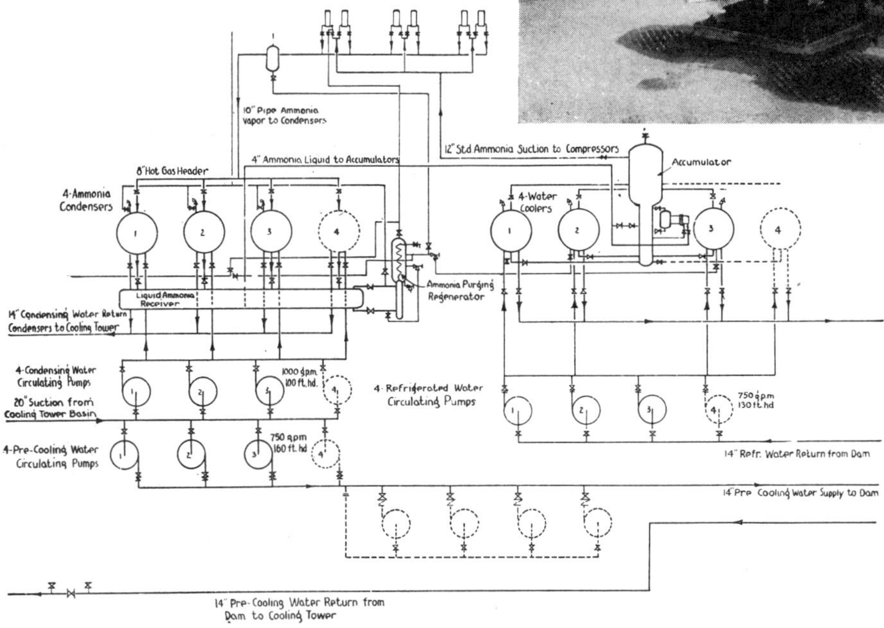


FLOW SHEET OF THE CONCRETE COOLING PLANT.

composition and kind of cement used and upon the type and density of the aggregates entering into the concrete. To remove from the mass the 40° imparted to it by setting would therefore require the extraction of approximately 38,600 Btu's for each cubic yard of the 3,325,000 cubic yards in the dam.

At the time bids for construction were taken it was planned to cool the concrete by circulating through it refrigerated water by means of successive layers of 2- or 2½-inch pipes. These pipes were to be spaced on approximately 10-foot centers, arranged hexagonally in the concrete, and were to form series of loops extending from an 8-foot slot in the center of the dam out to the junction of the concrete and rock on either side. These coils were to be supplied with refrigerated water from headers in the central slot at their respective levels.

Following the issuance of the original specifications, the investigations as to the methods to be used in cooling the concrete were continued at the Owyhee Dam, which was then under construction. These later findings, together with certain changes in the form and arrangement of Hoover Dam, led to modifications of the plan previously outlined. It was decided to use 1-inch OD cooling tubes placed in horizontal layers on approximately 5-foot centers, spaced hexagonally. Under this plan approximately 3,500,000 feet—662 miles—of tubing will be embedded in the dam.

The tentative program provides that the concrete shall be cooled to a temperature of 45°F. at the upstream face of the dam and to approximately 65° at the downstream face, with a progressive range between these two points for the internal body of the structure.

The Bureau of Reclamation specified only the nature of the piping to be placed in the dam and left to the contractors' option, subject to the Bureau's approval, the design and arrangement of the cooling plant and piping required. The original specifications called for a cooling plant having a capacity of approximately 600 tons of refrigeration, sufficient to deliver 2,100 gpm. of water at the dam and to cool it through a range of from 47° to 40°F. As the plans developed, it became necessary to make provision for the additional requirement of 3,000 gpm. of precooled water to be taken from the river or from any other suitable source, as the contractors elected. This water is to be circulated through the cooling pipes before refrigerated water is introduced, and will suffice to cool the concrete through half the range from its initial temperature to the final temperature specified. Refrigerated water will then be turned into the same coils to complete the cooling process.

In view of these modifications, it devolved upon the contractors to provide a precooling-water system and a refrigerated-water system each of which could be operated independently and yet be made to combine their common parts so as to effect maximum economy in their construction. After studying cost estimates and analyses of a number of systems of the two classes, one of each was selected and final designs were prepared.

The precooling water, which is taken from the river, falls over a cooling tower, flows to the plant, is pumped through a supply line to the dam, circulates through the header and coil systems for cooling the concrete, and returns to the top of the cooling tower where the heat that it has taken up is extracted from it. It is then collected in a basin at the foot of the tower to begin another cycle.

The refrigerated water circulates in a closed system. It is cooled by the action of ammonia, is pumped through supply lines into headers supplying the loops in the concrete, and then returns to the plant for extraction of the heat taken up, after which it is recirculated.

Refrigeration is effected by an ammonia compression system similar in most details to those used in making ice. As is well known, ammonia boils or evaporates at a low temperature under a low pressure and at a correspondingly higher temperature as the pressure is increased. By regulating its pressure it is, therefore, possible to cause it to boil or evaporate at any desired temperature and to obtain any cooling effect wanted within the limits of its thermal characteristics. In making ice the pressure is varied so that the ammonia will boil at a temperature below the freezing point of water. In this case, however, it is not required that freezing temperatures be attained, and the compression system is, accordingly, arranged so that the ammonia will boil at a temperature of 37° F., which is sufficient to cool the circulating water to the desired point.

Ammonia under a pressure high enough to hold it in the liquid state is allowed to expand through a valve which lowers its pressure to

approximately 50 pounds to the square inch, which is sufficiently low to cause it to evaporate at a temperature of 37° F. In changing to the vaporous state it takes up enormous quantities of heat. Evaporation is caused to take place in a vessel through which the water returning from the dam at a temperature of about 47° is circulated at a rate just fast enough to permit the removal of enough of its contained heat to lower its temperature to 40°. After exerting its cooling effect the ammonia vapor is compressed and cooled sufficiently to liquefy it, ready to be expanded again.

The ammonia-compression equipment consists of four Ingersoll-Rand machines which were originally a part of the air-compression plant installed to supply air for operating rock drills and other pneumatic equipment during the early stages of the construction work. Six Companies Inc. selected these units with the idea of later transforming them into ammonia compressors, and this has now been done by substituting suitable compression cylinders for those previously used. Four units were thus converted, but one of them has so far not been put in operation. Each machine has two 10&10x14 single-stage cylinders. The compressors are of the direct-driven electric type, and were originally equipped with over-size synchronous motors of 250 hp. capacity to meet the added power demands following their conversion to ammonia machines. Each compressor has a normal rated capacity of 275 tons of refrigeration.

In these units the ammonia vapors resulting from the expansion are compressed to approximately 190 pounds pressure. Their discharge temperature is around 200° F., and in order to liquefy them it is necessary to cool them. This is done in a Braun shell-and-tube-type condenser equipped with Babcock & Wilcox tubes. There are three of these condensers, one for each compressor. Each contains 430 one-inch OD seamless steel tubes 20 feet long. Combined, they provide approximately 2,250 square feet of surface. Cold water circulates through the tubes in four passes, thereby cooling and liquefying the ammonia vapors which are circulated around the tubes. Each of these condensers weighs about 8 tons. Their shells are of welded steel plate with welded integral tube sheets in which the tubes are expanded.

The liquid ammonia passes from the condensers to a welded steel receiver 30 inches in diameter and 17 feet long. Beside the receiver is a Gay purging regenerator for removing in one operation noncondensible gases from the system and also oil originating in the compressors. This is accomplished with the aid of ammonia from a high-pressure oil trap, which will be mentioned a little later. The evaporation of this ammonia condenses out and recovers the ammonia which is mixed with the noncondensible gases drawn from the various purging connections with the condensers, coolers, and accumulator. The piping is so arranged that the regenerator may be used for pumping out and evacuating any particular section or unit of the equipment upon which it is desired to work.

The condensing water is obtained from the same cooling tower used for the precooling-water system. This tower has a capacity to cool 6,000 gallons of water per minute through 75° F. to within 5° of the wet-bulb temperature specified at 75° with a wind of five miles per hour. It is built of redwood, and is 143 feet long, 16 feet wide, and 43 feet high. The water is distributed at its top by a 10-inch, redwood, stave header which has spray nozzles at 1-foot intervals. The tower was supplied by the Foster-Wheeler Company through the Consolidated Steel Company of Los Angeles, and was erected by Six Companies Inc. The lower cofferdam, on which the tower is located, will be removed after Hoover Dam has reached a height sufficient to permit storage to begin behind it. This will necessitate dismantling the tower; and it is expected that precooling and condensing water will thereafter be drawn from the reservoir.

Just before the ammonia vapors enter the header which supplies the condensers they pass through a high-pressure trap which captures particles of oil that may have passed over from the compressors and that would coat the heat-exchange surfaces and reduce their efficiency if not removed. This is a cylinder, about 18 inches in diameter and 4 feet high, which has a drain connection at the bottom and a purge connection at the top, where there is also a baffle set transversely to the flow of the ammonia vapor.

The exchange of heat between the return water from the dam and the ammonia takes place in three water coolers. These, like the condensers, are of the shell-and-tube type. The shell is a welded steel cylinder 30 inches in diameter and 20 feet long. Each cooler contains 375 one-inch OD stainless-steel tubes through which the water passes while the ammonia evaporates around them. The tube sheets are integral with the shell and are of heavy 2-inch construction to enable them to resist the pressure of the 600-foot head of water that will be met when the concrete in the top portion of the dam is being cooled.

The liquid ammonia is supplied to the coolers by a reservoir known as the accumulator. Six-inch ammonia-vapor return lines extend from the coolers to the accumulator to permit entrained particles of liquid ammonia to return to the vessel. The gaseous ammonia then goes to the compressors through a suction line which is taken off one side and near the top of the accumulator. To allow ample vapor-disengaging space, the accumulator is larger at the top than at the bottom and its upper 7-foot section is 4 feet in diameter while its lower 7-foot section is only 18 inches in diameter. On one side and in its lower part is a connection for a York 2-inch float control valve which automatically regulates the liquid level in the accumulator and coolers. There is a 4-inch liquid-ammonia inlet from a hand-operated expansion valve and a 2-inch inlet from the float control valve.

All-welded piping is used for the plant am-

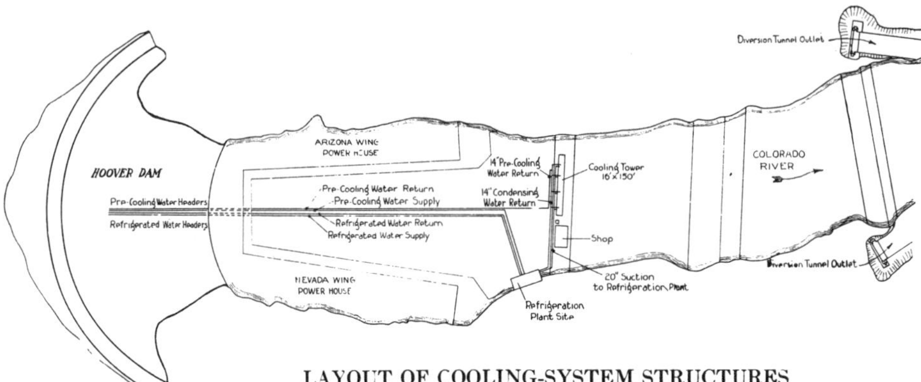


LAYOUT OF COOLING-SYSTEM STRUCTURES

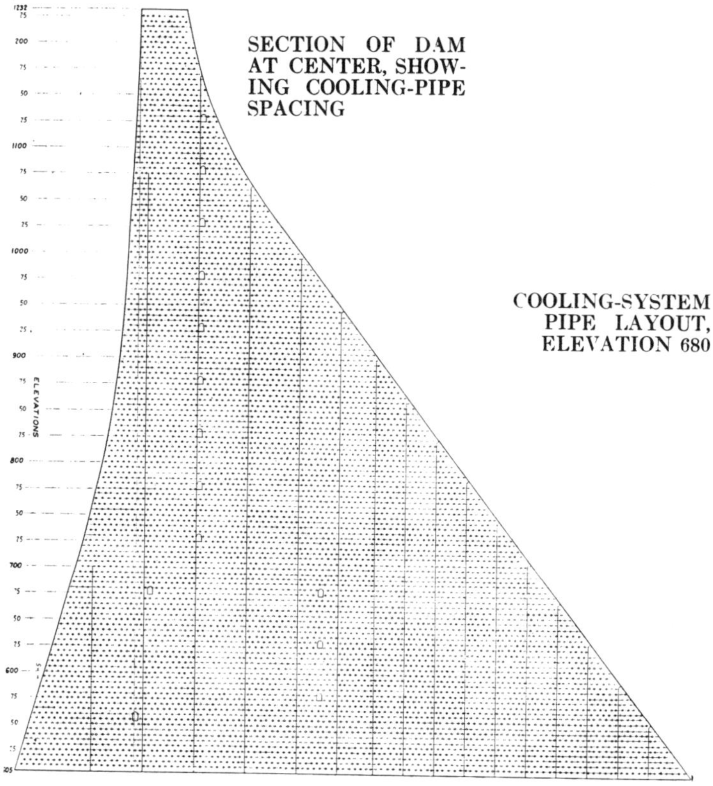

SECTION OF DAM AT CENTER, SHOWING COOLING-PIPE SPACING

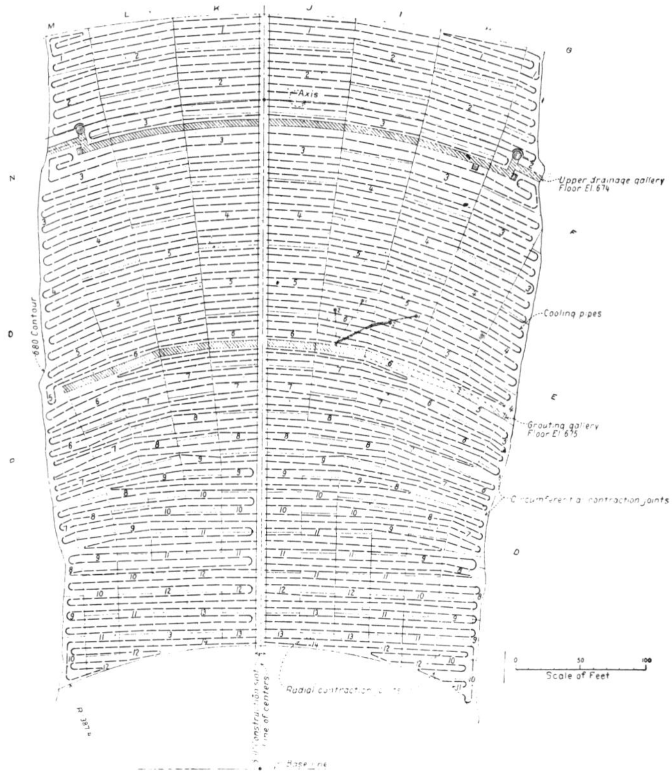


COOLING-SYSTEM PIPE LAYOUT, ELEVATION 680

monia system, flanged as required to accommodate fittings and connections to vessels. All lines, valves, fittings, and apparatus containing water or ammonia at low temperatures are insulated with 2 inches of cork surrounded by finished plastering and then painted to give a neat appearance.

It is anticipated that approximately 23 months will be required to complete the pouring of the concrete, during all of which period cooling water will be circulated through one or more of the various levels as the huge structure grows higher. Failures in the circulatory system would cause costly delays in the contractors' schedule, and it is, accordingly, of vital importance that the pumping equipment be both efficient and reliable. There are three sets of pumps, each of which performs an individual service. The first set —the precooling-water pumps—consists of four Cameron No. 4 ANV units, each of which has a capacity of 750 gpm. against 160 feet of head. They operate at 1,750 rpm., and are each driven by a Westinghouse 50-hp. motor. These pumps are intended to supply precooling water until the dam reaches elevation 750, which will be about one-third of its ultimate height. For service above that level will be installed booster pumps to operate in conjunction with the present pumps. They will serve to circulate the precooling water up to elevation 950. These booster units will each have a capacity of 750 gpm. against 200 feet of head, and will be designed for case and gland working pressures of 300 pounds. At the time they are being installed, the impellers of the present pumps will be changed to increase their head capacities to 200 feet.

Another set of pumps circulates the refrigerated water through the refrigerating plant and dam cycle. There are four of these, of which three are operated and one serves as a spare. These are Cameron No. 4 BNV units, built with extra-heavy casings to withstand 400 pounds pressure at the shut-off. Each pump has a rated capacity of 750 gpm. against 130 feet of head, and is operated at 1,750 rpm. by a Westinghouse 40-hp. motor.

The third group of pumps handles the water used to condense the ammonia vapors. These are four Cameron No. 5 LV units, each with a rated capacity of 1,000 gpm. against 100 feet of head and driven at 1,750 rpm. by a Westinghouse 40-hp. motor. Three are used at a time, the fourth being held in reserve.

A considerable length of large-diameter pipe is required to connect the refrigerating plant, the dam, and the cooling tower. A 20-inch line. 220 feet long, carries precooling and condensing water from the tower to the plant. Precooling water goes to the dam through a 14-inch line and is returned through another of the same size to the base of the cooling tower where it passes through two 10-inch risers to the distributing header at the top. Two 14-inch lines between the plant and the dam provide for the supply and return of the refrigerated water. The refrigerated-water lines are insulated with 2 inches of cork to reduce temperature rises in summer and to guard against freezing in winter. All these major lines are buried and backfilled with fine sand from the river bottom.

The U-shaped power plant will occupy the space just below the dam, with wings extending along each canyon wall. To minimize the difficulties of carrying the four 14-inch water lines up the face of the dam through the powerhouse construction it was decided to place them in the slot where they would be out of the way. When the cooling of the lower portion of the dam is completed, and the slot is concreted, these lines will remain embedded. Connections from them will be run out to the edge of the downstream face above the roof line of the power house, and from that point upward they will be carried along the face. The arrangement of this piping is illustrated.

To distribute the water from the 14-inch supply mains to the 1-inch cooling coils, a 6-inch header is run through the slot at either side, there being two supply headers for precooling water and two for refrigerated water. Connections to the inlet ends of the various cooling coils are taken off the headers by means of piping and 1-inch hose connections. Similar connections from the outlet ends of the coils to another pair of 6-inch headers are made for the return water. The 6-inch supply headers are taken off a common 8-inch cross header from each 14-inch supply line, and these are so manifolded that it is possible to supply either precooling water or refrigerated water by operating control valves. Similar 8-inch cross headers receive the return flow of the 6-inch headers and direct it to the respective return systems by means of valves.

As concreting progresses upward, sets of 6-inch headers are installed at various levels, as required by the number of cooling coils to be operated. In that portion of the dam midway between the bottom and the top, between approximate elevations 600 and 900, a set of these headers will be needed every 10 feet. An initial installation of approximately ten sets of headers will be required to supply water to all the coils which will be operated simultaneously. The lower four sets will be carrying refrigerated water and the upper six precooled water. As cooling by means of each level of coils is completed, the headers will be dismantled and reinstalled above the top set of headers then operating so as to provide cooling for the newest level of coils. As soon as each installation is made, precooling water will be started circulating. After precooling has been completed as specified, the valves at the manifolds from the 14-inch supply and return lines will be changed so as to shut off the precooled water and turn on the refrigerated water. This cycle will be repeated throughout the height of the dam, and each set of headers will be reinstalled approximately six times as the cooling is carried on at progressively higher levels.

The fabrication of the approximately 20,000 lineal feet of 6-inch headers that will be required is being carried on at a pipe shop on the lower cofferdam. All the connections for the supply and return lines for the cooling coils are welded on there. The headers are covered with 2 inches of cork insulation, and the pipe is trucked in 20-foot lengths to the dam and there set in place by cableway or by such other method as seems expedient for that level. These headers are supported in the slot by steel brackets made fast to the bolt anchors which have previously been used to hold the concrete forms in position. A walkway is provided at each level for placing these pipes, and a roof is maintained above the top level operating or being installed to protect the workmen from falling objects.

A tentative procedure of cooling operations, as specified by the Bureau of Reclamation, has been set up and used, but it is subject to such variations as prove expedient during the progress of the cooling. At present the control procedure embraces the following steps: obtaining starting temperatures of the concrete; adjusting the flow of water through the pipes; checking the progress of cooling; and checking the completion of cooling.

The Bureau of Reclamation has provided several methods for measuring the temperatures of the concrete and of the water entering and leaving the coils. There are being placed in the concrete, at the time of pouring and at a number of points in different levels, a series of thermocouples and resistance thermometers for determining directly the temperature of the concrete. The Bureau also has prepared a series of charts to enable its inspectors to determine, by a very simple calculation, the progress of cooling by measuring the water entering and leaving the cooling loops. Several variables are involved in determining the time required for cooling, namely: number of days cooling has proceeded; initial temperature of the concrete; present temperature of the concrete; the temperature difference between the water entering and leaving the coils; the number of days required to complete cooling; and the final desired temperature of the concrete. When certain of these variables are known, either specified or from field data, it is possible by using the charts provided by the Bureau to determine the values of the unknown variables in controlling the cooling. Starting concrete temperatures are taken before cooling begins by means of resistance thermometers embedded in the concrete at points between the embedded cooling pipes and also by means of resistance thermometers inserted various distances into the cooling pipes. This determination is being made for all coil zones prior to starting cooling.

The Bureau of Reclamation has determined that 4 gpm. is the most desirable rate of flow through the cooling coils. Inasmuch as the loops are of various lengths, the flow through each of the various coils from a set of headers must be regulated to compensate for these differences. A connection has been provided for the insertion of a meter in each loop for the direct measurement of the flow. As soon as a level of coils has been placed in operation, the loops are all regulated so that water at the rate of 4 gpm. flows through each.

As cooling progresses, measurements are taken to check the progress of the cooling. This may be done by drawing water samples into thermos flasks through small pet cocks provided at the inlet and outlet of each coil—by taking temperatures directly by inserting a thermometer in each sample and calculating the mean temperature of the concrete from these data. As an alternative, the flow through any loop may be stopped for 48 hours to allow the temperature to equalize in the concrete so that the water lying in the coil assumes the mean concrete temperature, which is then read by a thermometer immersed in the water. The completion of the cooling in a level of coils is checked both by direct measurement of the concrete temperature by means of the embedded thermometers and by measuring the water temperatures in and out of the coils. As soon as the cooling in any particular loop has been brought to the desired final temperature, circulation through the loop is stopped. The quantity of water supplied by the plant to the dam for both precooling and refrigeration is determined by the number of loops operating upon each circuit, since each coil requires a flow of 4 gpm.

WHERE THE HEAT ESCAPES

**Walkway at one level of the 8-foot slot through the center of the dam. The two pipes at either side are 6-inch insulated headers for supplying and returning the water which absorbs the heat generated by the setting of the cement. This is the first set of headers installed and it serves all cooling-pipe coils between elevations 525 and 575. Hose connections to some of these loops can be seen. The slot is roofed to prevent injury to workmen from falling objects.**

Grouting of the radial and circumferential contraction joints is begun as soon as cooling has been completed in a 50-foot grouting lift, and all the headers and staging are removed from the slot within the elevations cooled. The 8-foot slot is filled with concrete while grouting is proceeding. This operation of grouting and cooling makes the mass of the dam in each grouting lift a unified body of concrete with all temperature strains positively cared for. When this consolidating grouting, following the cooling, has been completed, the designer has a positive assurance that the mass of the dam will act exactly as intended in resisting the load imposed upon the structure by the pressure of the water impounded behind it.

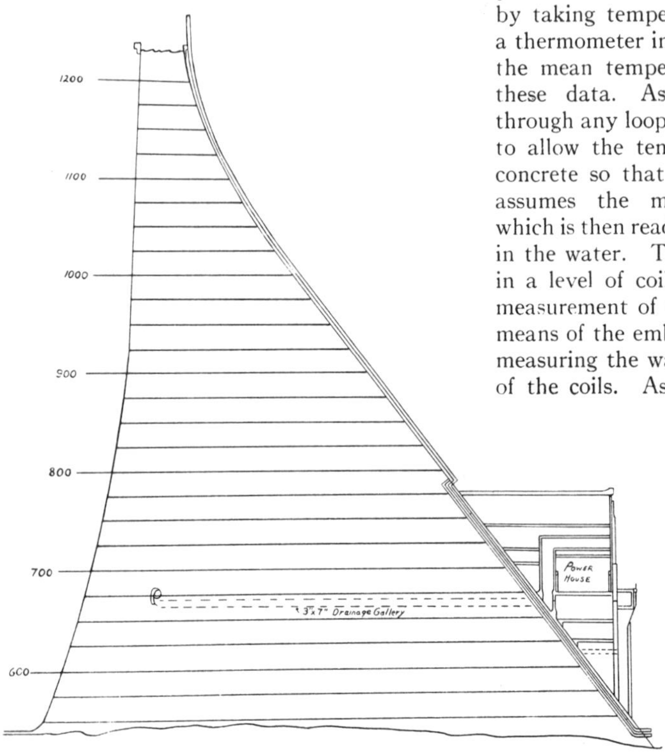


SUPPLY AND RETURN LINES

**During the pouring of the lower portion of the dam, the 14-inch precooling- and refrigerated-water mains will be carried upward in a slot so as not to interfere with the building of the central portion of the power house. Above the roof of that structure they will be run up the downstream face of the dam.**

SIXTEEN TONS OF CONCRETE

THE bucket shown above, containing 8 cubic yards of concrete, is being landed in one of the forms. The men closest to it on either side have tripped the safety latches on its bottom-dump doors and are getting into the clear as the cableway operator, more than 600 feet above them, releases the concrete upon signal from a man at the form. The form is 5 feet deep and about 50x60 feet on the sides. Its left side is being filled first, creating a sloping surface towards the right. This procedure is followed during hot weather to keep the surface of the concrete "live". At other times the form is filled progressively throughout. On the right a cableway is moving off with a bucket of concrete just taken from a car on the high-level railroad.

All photographs not otherwise credited are from Six Companies Inc.

# Construction of the Boulder Dam*

## Description of the Methods of Pouring the Concrete

LAWRENCE P. SOWLES†

PLACING CONCRETE BY CABLEWAY SYSTEM

This view, looking upstream at the dam from the lower catwalk across the canyon, is interesting because it shows the relationship between the pouring area and the overhead cableway system. In the upper part of the picture are shown the carriages of all five cableways. Suspended from them by fall lines are 8-cubic-yard buckets of concrete. The concrete delivery railroad from the Lomix plant is seen along the edge of the Nevada canyon wall at the left. The structure across the canyon in the middle of the picture is a catwalk for workmen. For a plan drawing of the cableway system see page 37.

THE construction of Boulder Dam is primarily interesting because of the enormity of the work involved. The building of a big concrete dam consists essentially of placing a large volume of concrete in one mass in accordance with a specific design, and of providing the necessary appurtenances and accessories. Except for their scale of application, the operations in the case of Boulder Dam are much the same as those that have been performed in rearing many other dams of more or less consequence.

Owing to the unprecedented size of this particular structure, unusual specifications and stipulations for controlling the work were set down by the U. S. Bureau of Reclamation to insure an enduring and safe dam. To satisfy these restrictions and to handle the immense quantities of concrete involved at a rate never before attained, Six Companies Inc. have organized their forces and equipment to bring about a concentration of activities that stamps this undertaking as unique among all similar ones that have gone before it. It is not easy to grasp what it means to mix, transport, and place 3,220,000 cubic yards of concrete in a tapering mass 727 feet high; to deposit it in several hundred individual columns in conformance with rigid rules; to provide, the while, for galleries, drainage structures, and grouting and cooling pipes; to do all these things in an orderly, painstaking manner; and to maintain a schedule that will assure the completion of this phase of the job within the next two years. Perhaps it will help us to comprehend the huge scope of the activities if we state that 25 carloads of cement and approximately 150 carloads of aggregates are required to make up the concrete that is placed every average day.

The design of the dam is based upon researches which have been carried on almost since the Bureau of Reclamation was created in 1902, and upon the accumulated experience obtained in building more than 50 concrete dams during that period. Incidentally, Boulder Dam will contain more concrete than the preceding 50, a circumstance that warranted special intensive study to determine beforehand how concrete would behave when poured in a mass of this size and height, and to make it possible to specify materials, methods, and

†Engineering Department, Six Companies Inc.

*Seventeenth of a series of articles on the Colorado River and the building of the Boulder (formerly Hoover) Dam.

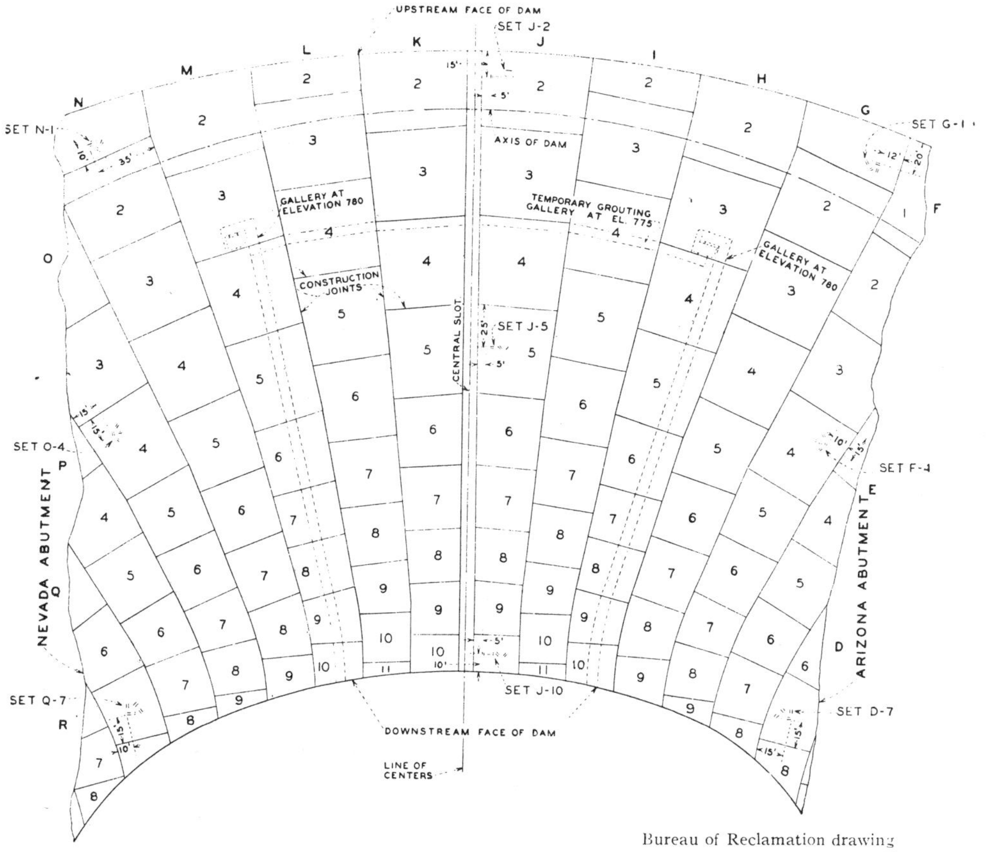


Bureau of Reclamation drawing

**PATTERN OF THE DAM**

Boulder Dam is being built of regimented columns lettered as to rows and numbered as to their positions. Owing to the taper of the mass from bottom to top and the variable width of the canyon, the pattern of the dam is an ever-changing design. At the left is a plan of the columns as they are disposed at elevation 800, a little below mid-height of the structure. The notations around the edges, such as "Set O-4," refer to the locations of strain meters to be embedded in the concrete. Below is shown the dam as it actually appears at elevation approximately 800.

practices that would serve to best advantage.

Space will not permit going into the details of the many and interesting tests that were conducted, or to show how they influenced the plans and specifications. These matters were summed up recently in a published statement by B. W. Steele, engineer on design of dams for the bureau. Mr. Steele stated that it has been the aim of the designers to produce a mass concrete of greater durability and higher quality than heretofore placed by the bureau, and that this purpose guided them in preparing specifications for the concrete plants, for the mix for the mass concrete, for the blended low-heat cement of a controlled composition, for bottom-dump concrete buckets, and in making provisions relative to the placement and curing of mass concrete. Thus far, the experience in connection with the actual construction of the dam indicates that these efforts have been entirely successful and have led to the following conclusions: That the structural qualities and the graduation of the aggregates are the best ever obtained in a dam built by the Bureau of Reclamation; that the contractors' plant for the production of concrete is probably superior to any ever used on similar work; that the strength of the concrete as poured is ample for any stress condition that may result in the dam; that a water-cement ratio of 0.53, using one barrel of cement to each yard of concrete with a slump of 3 inches at the forms, makes for a more uniform concrete than can be obtained with a lower slump under the conditions encountered on the job; that uniformity is the most important factor in the entire concrete-manufacturing process; and that all specifications covering concrete must be considered with regard to their effect on uniformity.

Concrete entering into the dam is produced at two locations: at Lomix, which is situated on a bench on the Nevada side of the canyon bottom and about 4,000 feet upstream from the dam site, and at Himix, which occupies a site on top of the Nevada rim of the canyon. In either case, the concrete is moved from the mixing plants in 8-cubic-yard bottom-dump buckets placed on railroad cars; is spotted in position underneath any one of the five cableways that span the canyon; and is conveyed from the cars to the forms by cableway. A minor variation in this general procedure will be described later.

The railroad which delivers concrete from the Lomix plant to the dam runs at elevation 720, and all the concrete placed in the dam up to this level came from that plant. Pouring is now in progress at elevations above 720, and from there on to elevation 975 concrete will be furnished from both plants. When elevation 975 or thereabouts is reached, the Lomix plant will, it is contemplated, cease to operate. From elevation 975 to the crest, at elevation 1232, concrete will be supplied from the Himix plant exclusively. For the time being we shall confine ourselves to an account of the methods followed in placing concrete produced at Lomix and in that portion of the dam below elevation 720.

The railroad from the Lomix plant to the dam site is a part of the line designated as the Canyon Railroad. It was used extensively during the early building period for removing muck from excavation areas. It is of standard-gauge, double-track construction, and hauling over it was done with steam locomotives until concrete transporting began, when the section from Lomix to the dam site was converted to electric operation. Third rails were provided except at the mixing plant, at the downstream terminus where the buckets of concrete were handled by cableway, and at switches. They were omitted at those points to avoid the possibility of men or water coming in contact with them.

The locomotives are of the storage-battery type—the batteries furnishing the required operating power where there is no third rail and being charged from the third rail at a rate that makes it unnecessary to remove the units from service for recharging. The third rails are supplied with 320-volt direct current by two generators located midway of their length. The locomotives originally were of 36-inch gauge and used for hauling muck in the 12x12 top headings of the diversion tunnels. They were converted to standard gauge and were equipped with air brakes, air compressors, and third-rail shoes in the machine shops of Six Companies Inc.

For concrete haulage, a special car was built on a standard flat-car bed. A back wall and five lateral partition walls were installed, thereby creating four wells each approximately 10 feet across and designed to accommodate a concrete bucket. The side of the car towards the dam was left open to facilitate the landing and removal of the buckets. The tops of the back wall and partitions were made wide enough to provide walkways for the men who hook and unhook buckets at the cableway

LOOKING THE DAM IN THE FACE

**The downstream face of the dam as it appeared on January 24, 1934. The irregularity of the lower blocks is occasioned by the fact that they will be used as footings for the power house which will occupy this space. The cooling-pipe slot with the headers rising in it is shown in the center. The vertical structure just to the left of the slot is an elevator which provides access to and from the top of the dam. This picture gives a fair idea of the immense amount of incidental structural work that is required in connection with the placing of the concrete. During the pouring of the concrete, the labor force on the dam has reached the highest number employed since the work began.**

loading points. The spacing of the wells was a little more than half that of the four mixers at the Lomix plant. When loading, a car was spotted so that two buckets, in alternate wells, received half of their 8-cubic-yard charge from the first and second mixers in the line of four, and was then moved alongside mixers Nos. 3 and 4 from which they received the second half of their charge.

Each concrete train consisted of a locomotive and one car with two loaded buckets in alternate wells, as just described. Upon arriving at the terminus adjacent to the dam, the car was spotted so that one of the two empty wells was directly underneath the hook of a cableway bearing an empty bucket. This was then landed on the car. While the hookers were disengaging the slings from the rim of the bucket, the train moved on until the adjacent loaded bucket was directly beneath the cableway falls, so that it could be hooked on to the slings and hoisted and carried by the cableway to the block in the dam in which it was to be poured. The train then proceeded to a point where another empty was waiting suspended from a cableway. There the same procedure was repeated—the train moving back to Lomix with its two empty buckets properly spaced to correspond with the mixers.

The section of the railroad at the downstream or dam end was one of the most interesting and expensive stretches ever built in this country. The original terminus consisted of a timber trestle, 100 feet high in places and 160 feet long, that carried the tracks almost up to the dam site. As then constituted, the railroad was used for hauling supplies into the canyon construction zone and for disposing of much of the muck that came from the upstream portals of the diversion tunnels. Its extension within the dam lines had to await the completion of scaling operations on the Nevada canyon wall, which was to serve as the dam abutment on that side. Actually it was not extended until just prior to the beginning of concrete pouring in the dam area. Within the dam lines the construction consisted of steel bents and floor beams with timber stringers, ties, and flooring. Steel was used for the under supports because, as concreting progressed upward, these members had to be cast within the mass of the dam.

As the canyon wall beneath the trestle was practically vertical, extremely long trestle legs were required to support its overhanging edge, even though heavy cuts were made into the rock on the canyon-wall side to reduce this overhang. To facilitate the placing of concrete underneath the trestle, the bents were made as simple as possible. Instead of the usual longitudinal braces, the bents were tied together and made rigid by means of flooring laid diagonally both ways on top of the ties in the space between the rails. As the elevation of the concrete neared that of the top of the bents, the stringers were removed and another method of handling the concrete was adopted. This will be described later.

Upon reaching the trestle section, the con-

DELIVERING CONCRETE FROM LOMIX

A double-track railroad, clinging to the sheer wall of the canyon with the support of long steel trestle legs, made it possible to deliver concrete for the lower section of the dam directly alongside of it. The 8-cubic-yard buckets, two to a car, were hauled from the Lomix plant by electric locomotives. As soon as a car was spotted at the dam site, a cableway deposited an empty bucket in one of the four wells and then picked up a full one for transportation to the forms. Repetition of this procedure disposed of the load and enabled the train to return to the mixing plant. In the view above, an empty bucket is being landed on the farthest car and a full one is being moved for pouring in a form near the lower left corner of the blocks. The second picture shows a concrete car being loaded at the Lomix plant. The buckets are spaced so that each is directly under the discharge of a 4-cubic-yard mixer.

Bureau of Reclamation photo

crete trains came under the control of a dispatcher who directed all their movements to and from the cableway spotting location. By means of a bank of levers in front of him, he maneuvered all switches leading to the trestle and for the crossovers on the trestle. The actual opening and closing of the switches were done with pistons actuated by compressed air. This system proved efficient in preventing tie-ups at one of the bottleneck points in the line.

The cableway system by which the concrete is placed in the dam is described on page 9. The method of handling the buckets with these aerial transportation agencies is one of the intensely interesting and spectacular features of the work at the dam. The difficulties attending the correct handling, landing, and spotting of a loaded bucket during the pouring of the lower part of the dam can be comprehended when it is realized that it constituted a weight of approximately 20 tons dangling at the lower end of 650-foot fall lines extending down from two carriages riding upon the track cable overhead.

For each cableway there was a signalman at the loading point and another at the pouring point. Both were connected by the same telephone circuit with the cableway operator in the head tower on the Nevada rim of the canyon far above them and out of their sight. Thus, they were able to give signals orally, as well as by the conventional system of bells and lights. By joint effort, these two signalmen had to direct the movements of the buckets between the delivery trains and the forms. Each loaded bucket had to be spotted precisely at the desired point for pouring and, when empty, had to be piloted unerringly back to its particular well on the car. These maneuvers involved not only cross-canyon movements but also longitudinal shiftings. The latter were guided solely by the signalman on the forms, his directions causing the cableway operator to propel the towers upstream or downstream on their runways, as need be. All other movements were under the joint control of the signalmen. Where one relinquished command, the other took it up.

It required no mean skill, alertness, and coordination for the signalmen to determine the exact point and time at which to stop the carriage on the track cable an eighth of a mile above so that the bucket might complete its swing immediately over the particular area in the block for which the concrete was destined and then, while the bucket was momentarily stationary at the end of its swing, to bring the carriages promptly into position directly above that point so as to catch and stop the pendulum action of the bucket and to prevent any back swing. This accomplished, the bucket would be lowered to the surface of the block where two men of the placing crew released the safety latches on the bottom of the bucket with their shovels. Meanwhile they guided the load to the spot where they wished the concrete poured, and then scurried into the clear as the signalman flashed word to the cableway operator to dump the 16-ton charge.

As soon as a bucket was empty, it was closed and hoisted to the required height for landing on the car on the trestle. The carriages overhead were next pulled toward the canyon wall, and after a lapse of several seconds, during which they had traveled as much as several hundred feet, the bucket began to swing and to attain its full velocity. Here again it was necessary for the signalman to gauge just how far to move the carriages so that the pendulum motion of the bucket would carry it to the exact point above the car on which it was to be landed, and then, while it was momentarily at rest, to bring the carriages quickly and directly above the bucket and to stop them there while the hook tender on the car assisted in the final spotting over the designated well.

Let it be said to the credit of the signalmen that there were but few times when the buckets were banged against the rock wall behind the loading point on the trestle, while there were thousands of instances when the buckets were landed neatly and smoothly on the cars. It was, indeed, fascinating to watch bucket after bucket come gliding up to the landing point, slide into a well on the car and be lowered into place—all with practically one motion which represented the ultimate in concerted action and timing. Especially careful were the signalmen in spotting buckets at the pour points on the columns to prevent them from hitting men or forms, which latter contingency might have proved extremely disastrous to all workmen in the vicinity.

As previously stated, concrete was delivered right from the Lomix plant to cableway loading points by the Canyon Railroad during the pouring of all that part of the dam from the lowermost point (elevation approximately 525) to the railroad level at elevation 720. Since the space occupied by that section of the railroad lying within the dam lines was itself to become filled with concrete, its use obviously had to end when pouring reached elevation 720. However, there were certain economies to be gained by continuing production of concrete at Lomix, and, accordingly, when the system was changed, this was provided for. Cableway 5 was still in a vertical plane with that portion of the 720 trestle immediately upstream from the dam, and was therefore able to load directly from the cars and to place the concrete by the traveling towers as required. To make it possible also to utilize Cableway 6, a 20-ton, 140-foot-boom stiff-leg derrick was installed in such a position as to enable it to transfer buckets from the cars on this trestle to adjacent points on the dam, from which the cableway picked them up. Cableways 7 and 8 were based on the Himix plant for their source of supply, receiving the buckets from cars on the high-level railroad, at elevation 1235, and lowering them to the working level on the dam. As they were more heavily loaded on the downward trip than on the return, their lowering speed was increased by reeving these cableways 3-part. This arrangement will suffice as long as it is economically practicable to continue the Lomix plant in production. After that, all the concrete will be supplied by the Himix plant.

MODIFIED CONCRETE DELIVERY SYSTEM

When the dam reached elevation 720, the section of the railroad within the limits of the structure had to be removed. This required terminating the line at the upstream face of the dam. Only one of the cableways could be shifted far enough upstream to permit it to load from that site. To make it possible to use a second cableway, a 20-ton, stiff-leg derrick was installed for transferring buckets from the railroad cars to the adjacent blocks of concrete. This picture was made on March 7, 1934, and shows the highest blocks at elevation approximately 790, or 285 feet above the lowest part of the foundation.

Acme photo

After the dam reached elevation 720, the full capacity of the Lomix plant was no longer required, so one mixer and one set of batchers were removed and reinstalled in the Himix plant. The three mixers then remaining at Lomix were arranged for feeding from one line of batchers. Before the Himix plant came into use for supplying dam concrete, it had been furnishing material for spillway structures and for penstock-tunnel linings. It had four mixers served by three lines of batchers. The additional equipment from Lomix gave it five mixers and four lines of batchers.

Eight-cubic-yard buckets were selected because they facilitated meeting the specifications which prohibited continuous pouring at the dam and stipulated that "a mixer batch, or combination of mixer batches, shall be conveyed in one mass to its location in the dam." The buckets are of welded-steel construction, cylindrical in shape, approximately 6 feet 10 inches in diameter, and 7 feet deep. Each is suspended from the cableway hoist and dump blocks by two sets of slings—one set being made fast to the shell of the bucket by sockets and removable pins and the other set being made fast by sockets and removable pins to the I-bars supporting the bottom-dump doors of the bucket. The possibility of accidentally dumping a bucket in mid-air is eliminated by safety latches on lugs at the points of the doors.

As previously mentioned, buckets of this type are generally used for pouring. However, where the dam abuts against the canyon wall, or where a block becomes very narrow because

WHERE THE DAM WILL REST

Excavation of the river bed revealed conditions there essentially as indicated by preliminary exploratory drilling. Bedrock on either side was struck at elevation approximately 600, but in the center there was an inner gorge that had been scoured to a further depth of 80 feet. This view shows pouring of the first block of the dam on June 6, 1933. The wagon drill in the center foreground was being used to drill grout holes, some of which were being filled by the pump mounted on the truck at the right. In the distance is the skeleton of the water cooling tower that now serves in connection with the cooling of the concrete after it is poured. The structure along the canyon wall at the right is the railroad, at elevation 720, over which concrete for the lower section of the dam was delivered.

of the backward slope at the dam face, the width of a form is sometimes not great enough to admit a bucket of such size. These fractional pours, as they are known, were at first made by lowering the concrete in 4-cubic-yard agitators. Later on, an 8-cubic-yard bucket, having a hand-controlled radial gate, was developed for this purpose and is now being employed.

The working of the concrete into the forms after it has been deposited is done mainly by men in rubber boots who walk around in the mass, spreading it into place along the edges to insure neat form finishes, and tamping it around pipes, galleries, and such other structures as have to be cast in the dam. As originally specified, the concrete was to be deposited in 1-foot layers extending over the entire form area. A trial of this method disclosed that it was impossible to maintain a "live" working surface during hot weather because the concrete dried too rapidly. It was, accordingly, decided to modify the procedure when weather conditions necessitated it by starting the pour at the downstream edge of each block, building up that side to the top of the form, and keeping the working face inclined toward the upstream edge of the dam—subsequent pours on the low side then progressively filling the form. This method is proving satisfactory, and keeps the working face "live" in even the largest blocks. During the winter months the procedure originally outlined is used. Vibrators are generally resorted to only where it is difficult to work the concrete by hand, as around embedded structures. A placing crew consists of seven men and a foreman in the larger blocks, and of five men and a foreman in the smaller ones.

A number of related operations must be carried on simultaneously with the placing of the concrete, and these must be timed and coordinated so as not to delay the pouring schedule. Among these are: placing drain tiles, grout stops, cooling pipes, gallery forms, grout pipes for contraction joints, thermometers, strain meters, and other instruments, as well as raising and checking forms. Pouring in each form is done in 5-foot lifts. As soon as a lift has set to a certain degree, a crew enters the form to clean up the surface of the concrete. This is done by scouring it with a jet of compressed air and high-pressure water, a method that produces a fine stream of water particles traveling at a high velocity and having an effect similar to that of a sand blast. This treatment removes any scum or laitance which may have formed, leaving a roughened surface with the aggregates exposed and insuring a good bond with the next pour of concrete. The water also serves to carry off the undesirable material and to convey it to a drain hole in the lowest block in the area.

Forms are generally left in place for about 24 hours after a pour. They are then stripped off, raised, and reset above in preparation for the next pour. When in position, they are checked for grade and line by Government engineers, who provide any additional points required to set them with the accuracy demanded by the specifications. The forms are hoisted by means of small A-frames on each of which is mounted a hand winch. Ordinarily, two frames are used for each panel, but three are needed to raise the longer ones, which have an extreme length of 61 feet.

The forms consist of wood panels lined with sheet metal, and are held in position by she-bolts, tie-rods, and braces which are attached to small hook bolts, called pigtails, embedded in the concrete. These hook bolts are placed

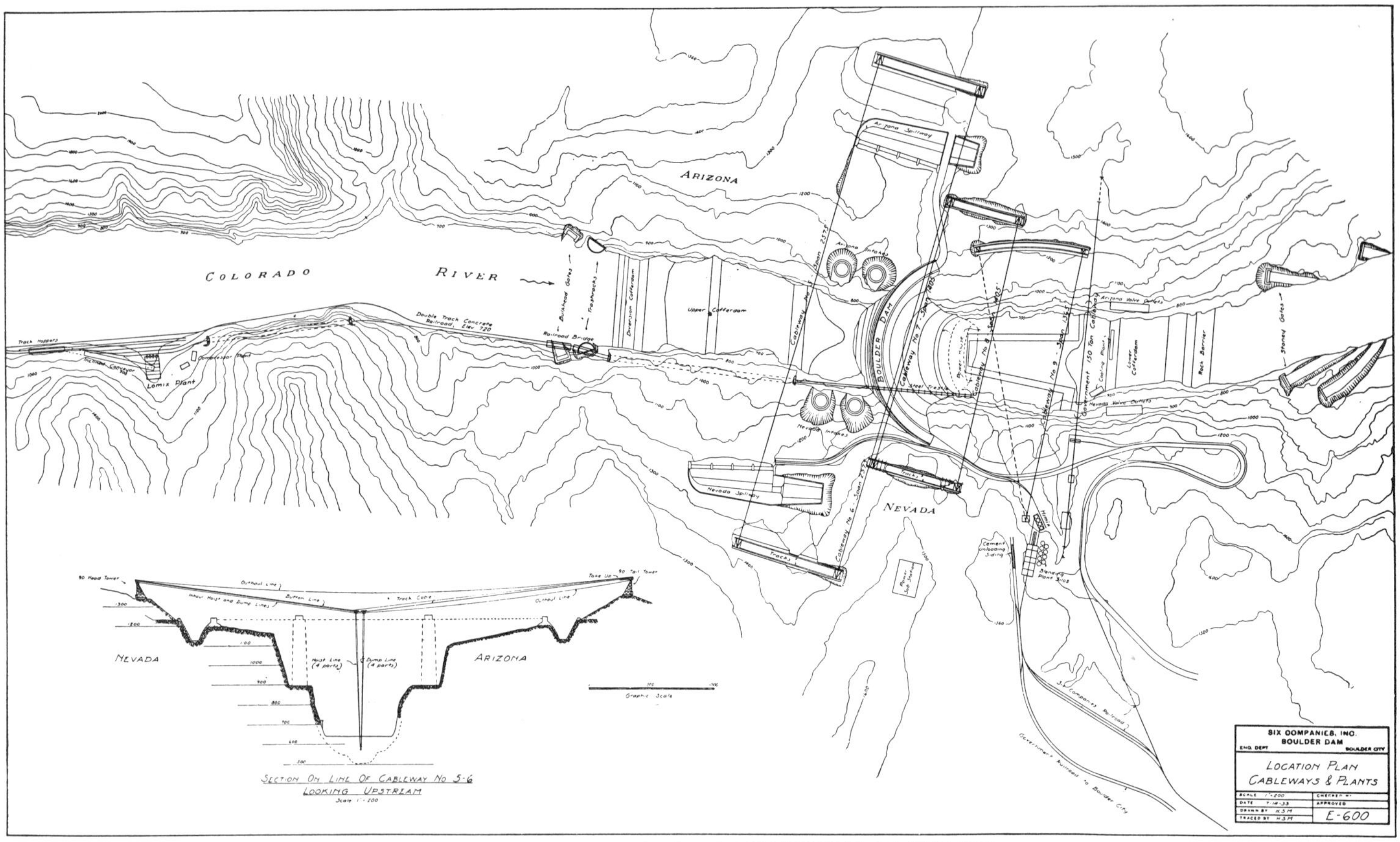


**LOCATION PLAN OF CONCRETE-HANDLING FACILITIES**

**All concrete poured in the dam from the lowest point at elevation 505 up to elevation 720 was produced at Lomix, shown near the left edge. The mixers discharged into 8-cubic-yard buckets placed on flat cars, and these were then hauled to the dam site, 4,000 feet downstream, where they were picked up by one of the five 20-ton aerial cableways that span the canyon high above and transported to pouring areas. Between elevations 720 and 975 the dam is being built with concrete from both the Lomix and Himix plants. The latter is located on the Nevada rim of the canyon, and from it a railroad runs at elevation 1235 underneath the five cableways. While the lower portion of the dam was being poured, the Himix plant was furnishing concrete for the spillways and intake towers on either side of the dam and for the lining of the penstock tunnels and other structures. It will be noted that the cableways, by means of their traveling towers, can deliver concrete to every section of the dam, power house, and appurtenant works. The Government 150-ton cableway, which appears as a straight line at the right of the contractors' cableway system, is being used for handling steel pipe for lining the penstock tunnels. Later it will also be utilized for lowering the generating equipment into the power house.**

in the top of each lift, at intervals of approximately 2½ feet, ready for holding the bottoms of the forms for the succeeding pour. Forms are raised a panel at a time; accurately lined at the top by means of braces and steel ties; checked for alignment; and then made tight.

At this point a pipe-fitting crew enters the block to place the grout piping and outlet boxes. The grouting-system tubing varies from 2 inches to ½ inch, outside diameter, depending upon its position. All couplings and tees are of the compression type and are similar to those used with the cooling pipe. The tubing carries the grout from headers to small units located at intervals along both radial and circumferential contraction joints. Half of the grouting boxes required for a contraction-joint section are placed with each of the two pours that are common to that joint. Incomplete portions of the system are carefully capped to protect the piping from damage and to prevent it from becoming filled with concrete and thereby destroying its usefulness for grouting purposes. The piping is arranged so that the radial contraction joints can be grouted separately from the circumferential joints in 50-foot lifts throughout the height of the dam. There are independent supply and return lines in each system so as to make sure that the fresh grout will circulate through the headers and then flow to the risers and through them to the boxes that distribute it in the contraction joints.

Divisions are required at various points to keep the grout in these two systems separate. In addition to a copper grout stop, which is used for this purpose in the interior of the dam structure, a Monel-metal water stop is provided at radial contraction joints at the upper face of the dam. There is also a similar stop at the downstream edges of radial joints below elevation 675, where tailrace water might reach the copper grout stop and corrode it. The specifications require that these stops shall be made absolutely grout tight by riveting and welding. Copper stops are brazed with acetylene torches and bronze rod. Monel-metal stops are welded by means of Westinghouse alternating-current welding machines and Monel rod. After being welded, the grout stops are covered with a layer of asphaltic emulsion which prevents leakage and too tight a bond between them and the concrete, thus allowing the small amount of slippage which is necessary when a joint opens owing to the contraction of the blocks of concrete on either side of it.

When the grouting system is in position, another crew appears to place the cooling-system piping. This consists of lengths of 1-inch outside diameter, 14-gauge, butt-welded tubing joined by means of compression-type couplings. These are laid on top of the concrete and, to prevent their shifting when the next pour is made, are fastened with wires which have been previously embedded in the surface for that purpose.

An ingenious method is used for protecting the exposed pipe-end couplings at the edges of blocks that are higher than adjacent ones. A small pressed-metal disk, similar to a 6-inch pie plate and having in its center a circular hole through which the end of the coupling is inserted, is nailed to the form with the depressed portion towards the inside of the block. This serves to make a small recess in the finished concrete and provides working space for affixing the coupling by which the end of the pipe extending into it is linked with the adjoining section when the abutting block reaches that level.

When connecting pipe is run in low blocks, a 4-inch plate is placed with its top nesting in the base of the 6-inch plate on the adjoining block, thereby insuring that the coupling will be housed in a space free from concrete and

Bureau of Reclamation photos

## DETAILS OF WORK IN THE FORMS

Each column in the dam is built up 5 feet at a time, with a specified interval between pours. Before fresh concrete is placed in a block, the hardened surface of the previous pour is thoroughly cleaned as an aid in securing a firm bond between the two lifts. A jet of water and compressed air (upper right) exerts a scouring action similar to that of a sand blast, and effectively removes foreign matter and the laitance which forms on new concrete. Sometimes the surface is also brushed with stiff brooms (upper left).

The tramp, tramp, tramp of rubber-booted laborers is the principal means of working the wet concrete in the forms (lower right). Long-handled shovels steady the men and also help in distributing the mix in keyways and around form edges. Compressed-air vibrators (lower left) are used to advantage in working the concrete around and under embedded structures.

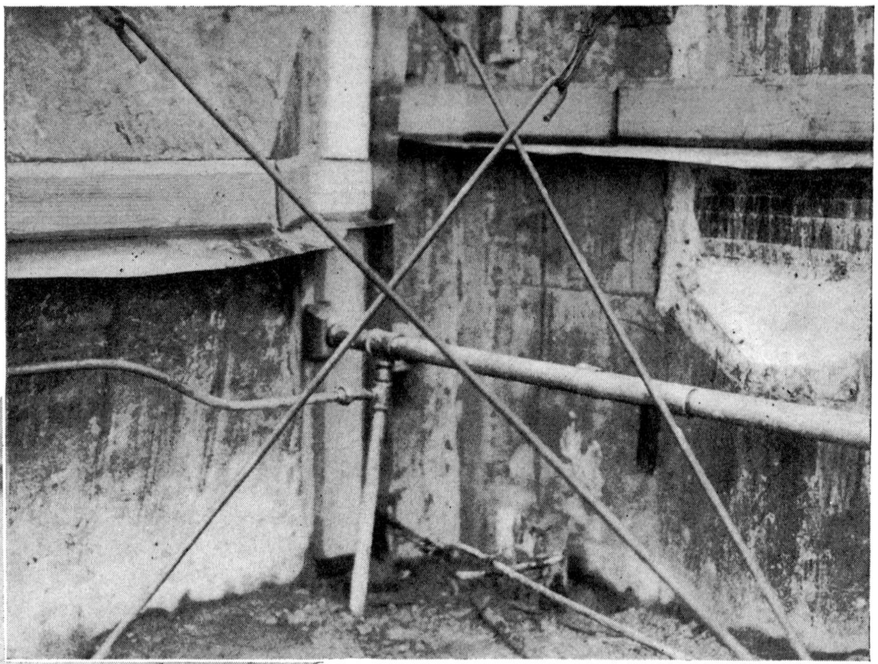

DETAILS OF GROUTING SYSTEM

Radial contraction joints are usual features in concrete dams, but the great size of Boulder Dam dictated the use also of circumferential contraction joints. The column arrangement was designed to produce a shrinkage crack adjacent to each block wide enough to permit satisfactory grouting without causing contraction fractures elsewhere. In order to grout these joints effectively after the concrete has been artificially cooled it is necessary to install piping and outlets and to leave open galleries from which the grout can be introduced. As the two sets of joints are grouted separately, copper stops must be placed in the concrete to eliminate all communication between them.

Bureau of Reclamation photos

will be able to act as an expansion joint. Before cooling pipe in a low block is connected with corresponding sections in higher blocks on either side, the flanking sections are blown out with high-pressure water to make certain that they are clear and unobstructed. If plugged lengths are found, the piping in that particular loop is rearranged, if practicable, to provide circulation for the cooling water through as large an area as possible.

Porous drain tiles of 8-inch inside diameter are placed at approximately 10-foot centers in vertical lines along the axis of the dam, extending from a drainage gallery near the base to the top of the structure. As the sections are 2½ feet long, two of them are needed for each drain in a 5-foot pour. This tile is made by Six Companies Inc. in a yard near Boulder City, and is of a special concrete containing, principally, aggregates from ⅛ to ⅜ inch and sufficient cement paste to bind the particles together while still providing a high porosity. The specifications require that each section shall by test pass 4 gallons of water per minute.

If the form being set up is adjacent to either of the canyon walls, the abutment rock is checked for soundness before any concrete is poured. Inspectors strike it with a hammer or heavy bar, and if it fails to give forth a clear, firm ring it is condemned as unsound and must be removed. As blasting is not permitted in such locations, the loose material is barred down by hand or gouged off with air-driven paving breakers. All the abutment areas were carefully gone over by high scalers during the preliminary stages of the construction, and most of the unsound surfaces that have been encountered since then have been small. Occasionally it is found that weathering has opened or deepened cracks in the rock that were begun by nature or that resulted from blasting during scaling operations. The material which is removed is allowed to fall down on the concrete surface below, where it is shoveled into a rock skip or bucket and moved by cableway to a clean-up car stationed on the railroad trestle, or to a dump truck.

Thermometers, strain meters, and other instruments which must be placed in the dam are inclosed in small receptacles that are embedded in the concrete. A small amount of wet concrete is worked around them by hand in such a manner that they will be protected from damage when pouring from the large buckets is started. These instrument installations require the setting of dowels and bolts and the running of considerable lengths of conduit to carry the connecting wires to the control boxes in the various galleries throughout the dam.

The setting up of forms for the extensive system of galleries constitutes another important accessory operation. One gallery runs along the abutments on either side of the canyon from about the lowest level of the dam clear to the top. It will be used for drainage, for foundation and abutment grouting, and for inspection of the abutment concrete. Two vertical galleries will serve as elevator shafts and will connect with all the horizontal galleries at various levels. Two galleries at elevation 705 will provide access to the power house from the lower elevator lobbies. Above these, and parallel to them, will be a number of radial galleries which will permit inspection of the concrete in the completed dam. There is also a series of galleries into which will be vented the headers for grouting the contraction joints. Wood forms of suitable shapes for the construction of these galleries are built in the contractors' lumber yard at Boulder City and assembled in sections in the various blocks where concrete is poured around them. The grouting galleries are filled with concrete after they have served their purpose and are, in turn, grouted to insure that no open joints remain.

When all the aforementioned auxiliary operations have been performed in a particular block, the surface is given a final clean-up and washing. The block is then inspected and checked by Bureau of Reclamation engineers, who certify as to its readiness to receive the next 5-foot lift of concrete. When approved, the block is posted on the pouring schedule. When the next pour is to be made, the placing crew enters the block accompanied by an inspector, who must be present during the deposition of all concrete. Preliminary to pouring the mass concrete there is applied a 1-inch layer of grout, consisting of sand, cement, and water. The purpose of this is to provide a firm bond between the hardened surface and the freshly mixed concrete.

Through the signalman on the form, the foreman of the placing crew orders the grout from the mixing plant. The bucket in which it arrives is stopped several feet above the form and then opened. The soupy grout strikes the surface with enough force to make it spread rapidly in a thin layer over the entire form area. With brooms and brushes, the crew then works the grout, thus bringing about its firm

EXTENSIVE RESEARCH PROGRAM

Because it was to be the greatest mass of concrete ever poured, Boulder Dam was studied intensively from every angle before construction began. Research in connection with the properties and behavior of concrete was carried on by the Bureau of Reclamation by means of special equipment placed in its Denver offices and also in collaboration with the Materials Testing Laboratory at the University of California where the concrete-strength testing machine shown above is located. Cement is secured from four mills and blended at the dam site to insure uniformity. Aggregates are specially treated and graded, and a high-quality concrete is obtained by using a water-cement content only 2 per cent greater than the void space in the dry rodded aggregates. This concrete, containing 3.96 cubic feet of cement per cubic yard, exhibits a compression strength of 3,200 pounds per square inch after 28 days.

contact with the previous surface and with installed structures. This grouting virtually eliminates the possibility, later, of water percolating through horizontal joints between pours—a circumstance that is believed to be responsible for considerable leakage through many concrete dams.

As described in the previous article, an 8-foot vertical slot extending through the center of the dam is left open in pouring the blocks to provide a means for circulating cooling water through the system of pipe coils. As soon as a 50-foot vertical section of the dam has cooled to the scheduled temperatures for that section, the slot within its limits is concreted. The lift is treated in ten successive stages of 5 feet each, with a minimum interval of 36 hours between pours. The concrete is placed progressively from the downstream face to the upstream face of the dam by means of special equipment.

At a level just above the upper limit of the 50-foot lift, standard-gauge rails are laid on stringers supported by 12x12-inch crossbeams wedged tightly at each end into notches cut in the concrete walls of the slot. An 8-cubic-yard car for delivering concrete travels on these rails. Underneath the track are two inverted rails which are hung from the track stringers. These rails, by means of four small trolleys, carry a small hopper and an "elephant trunk," which is a long, flexible pipe for conveying the concrete to the placing areas below. In operation, an 8-cubic-yard charge of concrete is delivered by cableway to the car at the upstream end of the slot. The car is then moved by an endless line, operated by a small hoist, to a point directly above the hopper on the elephant trunk. There the bottom gates are opened, and the concrete flows through the tubing to the desired location. With each successive 5-foot lift, the elephant trunk is shortened correspondingly.

During the excavating for the foundation of the dam, a number of springs were struck, and to dispose of their flow so that concrete placing might proceed, pipes were inserted into the cracks from which the water issued and the cracks were then leaded. The pipes were run in groups to manifolds, and from these headers the water was carried either to the cooling-pipe slot or to the nearest face of the dam. When the time came to concrete the lower part of the slot, it was necessary to transfer the discharge of a number of these drainage pipes to the face of the dam by running individual pipes from each spring system. The flow from some of the springs had ceased, from others it had increased; but it was required that the piping serving all of them be carried outside the dam limits to relieve the pressure on the foundation until grouting and drainage-hole drilling along the axis of the dam could be completed. During the first few pours of concrete in the slot, approximately 20,000 linear feet of small pipe of various diameters, that had been placed there, were embedded to serve as drains for these springs.

Prior to pouring concrete in the slot, the cooling pipes opening from either side of it are washed out and filled with neat-cement grout. Those that leak into the contraction joints are not grouted, inasmuch as grout might escape into the joints and plug some of the tubes of the grouting system inadvertently. The cooling-pipe coils thus left empty will fill with grout at the time the contraction joints are grouted.

After a 50-foot vertical section of the central slot has been concreted, grouting of the contraction joints in that lift is undertaken. The grout is pumped through the headers and distribution pipes to the outlet boxes on the face of each joint. About 50 square feet of circumferential joint area and about 30 square feet of radial joint are served by a box, and the number of boxes is apportioned accordingly.

Researches have indicated that it is possible to obtain a continuous film of grout, entirely filling the contraction joints, having a width of from 0.01 inch to the expected maximum of approximately $\frac{3}{16}$ inch. Special provisions govern the grouting operations. Only cement that passes through 200-mesh vibrating screens is permitted to be used; and, to prevent it from hydrating or forming lumps, storage for longer than ten days after screening is prohibited. It is specified that grout shall be pumped into radial and circumferential joints separately at such a rate as to fill the joint completely in not less than half an hour, and that it shall be circulated in a way so as to insure the removal of all air pockets from the system, to consolidate the grout film, and to press out the water into the concrete of the adjacent blocks. The pressure applied is determined by Bureau of Reclamation engineers, and is usually around 50 pounds per square inch at the grout pump.

Provision is also made for a film of grout between the concrete and rock abutments. The means for doing this consist of headers and risers leading to grout outlet boxes which are similar to those in the contraction joints except that half of the box is provided with a U-shaped iron wire which is grouted into a hole drilled a foot into the rock. This half of the box is supported by a small pad of concrete that holds it in position during the pouring of the adjacent block. The other half of the outlet is placed against the first one and connected to the supply system by compression couplings like those in the contraction joints. The actual grouting of these contacts

EXTREME BOTTOM OF DAM

The trench shown was at the upstream toe of the dam on the Nevada side of the central erosional gorge that the river had scoured in the canyon floor (see page 36). The bottom of this trench was at elevation approximately 505, and it was here that the lowermost concrete was poured. The network of pipes was installed to carry off the water from the numerous springs that were encountered. Grout will later be forced through these pipes to seal off the flow and to reduce the pressure on the foundation of the dam.

Bureau of Reclamation photo

proceeds in the same manner as in the case of the contraction joints.

Pressure grouting of the dam foundation is required to consolidate the rock and to fill existing cracks. Holes for introducing the grout are drilled in the bottom of the canyon and in the abutments to depths of from 50 to 150 feet. Pressures of from 100 to approximately 600 pounds per square inch will be used in applying the grout.

Supplying power and other utilities at the dam during all these operations is in itself a task of considerable magnitude and cost, since compressed air, water, light, and drainage facilities must be provided for every lift from the bottom of the structure to its full height of 727 feet. Compressed air is now produced at two plants having a combined capacity of about 13,000 cfm. At the height of the drilling program there were three plants with an aggregate capacity of 16,000 cfm. The two plants now in use are approximately two miles apart, both on the Nevada side of the river, and are connected by pipe line.

From the air main, branch lines are carried up the upstream face of the dam on brackets and up the downstream face in the slot which houses the cooling-water supply and return pipes. Branch air lines are also installed in the galleries, and from them risers, cast in the concrete, extend to the pouring surfaces.

CONCRETING CENTRAL SLOT

**The 8-foot central cooling slot is concreted in vertical lifts of 50 feet each after the dam sections on either side have been cooled. Concrete is delivered in buckets by cableway to the upstream face of the dam and there transferred to the car shown here. This car runs through the slot on tracks and discharges into a hopper from which flexible tubing extends downward to the pouring area. The 50-foot lift is divided into ten 5-foot pours, with a minimum of 36 hours between pours.**

From these principal delivery lines, laterals are run to various areas throughout the dam to supply manifolds having approximately twenty outlets to which hoses are attached as air is required for various purposes. Compressed air is primarily used for operating "Jackhamers" in drilling anchor holes for abutment grout boxes, anchor form holes, and light-bracket holes in the canyon walls; for running paving breakers in scaling down loose wall rock; for operating drills and other woodworking tools on the forms; and for the air-water jets that serve to clean the top surfaces of blocks.

Water supply and distribution must also keep pace with the concreting operations. Water for all constructional purposes is obtained from four wells sunk in the river bottom. There are two of these upstream from the dam and two on the downstream side. By means of manifolds and cross connections, water from any of these sources can be delivered almost anywhere on the job. In general, the upstream wells supply the Arizona and Nevada spillways high on the canyon walls above them, as well as the upper portion of the dam, and, during high-water stages of the river, form an emergency source of mixing water for the Lomix plant. The downstream wells supply the lower part of the dam, the penstock tunnels, and the valve-house benches, and furnish make-up water for the cooling tower of the refrigerated-water system.

A 6-inch water line parallels the 4-inch air main in the central slot up the downstream face of the dam. A water main also is carried up the upstream face of the dam. From these two risers extend branches to various areas and supply manifolds having approximately twelve hose connections. The principal uses of water are for curing the concrete and for cleaning operations, which have been described. Water for curing is applied in two ways: the vertical faces of concrete are kept moist by sprays issuing from ⅛-inch holes in a 1-inch pipe hung at the lower edges of the form panels. These pipes are moved upward when the panels are shifted, and require no attention from the waterman other than to see to it that the connections are made and that the flow of water is uninterrupted. Top surfaces of concrete blocks which are not being worked upon are kept continually moist by several men with hoses. Each man makes a complete circuit of the section of the dam assigned to him often enough so that no block will become completely dry before it is again sprinkled. The pumps deliver the water under sufficient head to provide a pressure of about 150 pounds per square inch at the working surfaces of the dam. This is required for cleaning, and is also desirable for sprinkling, as it enables a hoseman stationed on a high block to reach a number of surrounding lower blocks without the need of moving from block to block under difficulties.

Lights are essential to the successful pouring of concrete at the high rate that has been maintained thus far. Illumination has been so effectively provided that operations can be carried on as well by night as by day. Many auxiliary tasks such as laying pipes are done principally by the day shift, leaving the

FRACTIONAL FORM

**Where the concrete abuts against canyon walls, irregularly shaped blocks result. Areas that are too small to admit the bottom-dump buckets ordinarily used are poured by means of special buckets. In the triangular form against the wall can be seen the vertical risers for grouting the joint between the concrete and the rock. The form shows how vertical keyways (left) are produced for interlocking radial contraction joints. Grout riser pipes can be noticed at intervals along the near side of this form.**

swing and graveyard shifts to concentrate on pouring; and an examination of the records has disclosed that the night workers have generally placed more concrete per shift than the day crews.

Two types of lights are used: those that are semipermanent, and those that must be moved with each pour to give increased illumination in localized areas. The fixed lights are mounted on brackets on the canyon walls and are attached to the upper and lower catwalks across the canyon. These have large industrial reflectors, fitted with 1,000- or 1,500-watt globes, and are usually set in banks of three or six at each location. Local lights are of the portable flood-light or so-called dishpan type which has attracted wide attention among tourists. They are mounted on brackets or tripods and secured to any convenient structure in an area where men are working. Sets of small lights are also required inside the galleries where electricians make up the leads from the instruments to their respective terminal boards; along the several levels of the walkways in the central slot where the pipefitters make and change connections to the cooling-pipe headers; and at various points in the layout of blocks that is complicated on the downstream face of the dam by the power-house foundation.

Another important service is that of transportation for the workers. An inclined skip accommodating fifteen men is provided at the downstream face and two similar ones at the upstream face. These are railed platforms which are moved up and down the face of the dam on greased skids by 75-hp. motors. The hoist operator is in full view of the slides from top to bottom, and the workers give their signals directly to him. There is also a skipway extending from the rim of the Nevada abutment down to elevation 850. This aids the men in getting to and fro, and is also used for carrying materials. It has a capacity of 50 persons. Among the workers these elevators are known as "monkey slides." In order that it may not be necessary for the men to climb up and down ladders when going from one high block to another across an intervening low block, a light-weight foot bridge has been devised. One man at each end can handle one of these readily, and they are put in position while the forms are being raised.

The disposal of waste and the drainage of water from a series of high and low blocks is somewhat of a problem. Rock scaled from cliffs, lumber scraps, and debris of all sorts are cleared from high blocks and deposited on the lowest adjacent block, which becomes a catchall. At the collection point the waste is placed in a skip or bucket and removed by cableway.

Clean-up water is disposed of by means of a unique drainage system cast in the concrete of the dam. Water from high blocks is allowed to drain through holes in the forms to adjacent low blocks. Each of the latter opens at its lowermost point into a 4-inch vertical drain that is embedded in the concrete and connects with a 6-inch header that is almost horizontal. This carries the water to the nearest dam face. The openings into the drains are protected by conical screens to exclude particles of rock. This waste water is collected in a sump. If it settles clear, it is returned by a sump pump to the suction of the high-pressure pumps for re-use on the dam. Otherwise, it is pumped to a diversion tunnel that discharges into the river.

The coördination of these manifold operations and detailed tasks incidental to the actual pouring of the concrete is the real masterpiece of organization in connection with the contract. Principal credit for the smooth-working system which is resulting in record-breaking progress is due General Superintendent F. T. Crowe. Upon his shoulders the directors of Six Companies Inc. placed the chief responsibility for carrying out this huge undertaking. With the help of his able assistants he has set new concrete-placing speeds.

The program outlined by Mr. Crowe is being put into effect by A. H. Ayers, chief engineer, and by Bernard "Woody" Williams, assistant general superintendent, and their respective staffs. The field organization was selected with the same care that was exercised in the preparation of the construction plan. It is impossible here to mention more than a few of the keymen who are contributing to the successful outcome of this largest construction job ever undertaken in behalf of the Government.

Karl Collett, Frank Bryant, and Ed Wattis have charge, respectively, of the graveyard, day, and night shifts. Tom Price is superintendent of the gravel plant and the railroad, and in that capacity he oversees all operations having to do with the concrete aggregates from the time they are excavated from the gravel pit nine miles up the river until they are delivered ready for use to the two mixing plants. Dave Williams is in charge of the cableway operations and rigging. Ira Carpenter supervises the operation of the two concrete-mixing plants. The placing of pipe in the dam, the maintenance of air and water lines leading to the dam, and the water-supply system are directed by A. L. Reed, pipe foreman. The operator of the compressor plants and of the concrete-cooling-water system is Gus Larson. C. A. Harris, chief electrician, and his crew keep power on the line for the motors of the cableways, mixing plants, and pumps, and maintain the lighting system as well as the telephone system between signalmen and cableway operators. Cy Bous, master mechanic, and George L. Malan, machine-shop foreman, are responsible for keeping the equipment in condition for highly efficient work.

Such plant engineering as is required in connection with the contract—the designing, for example, of cableways, mixing plants, railroad trestles, derricks, forms, etc., etc.—has been done by the engineering department of Six Companies Inc. directed by Mr. Ayers and by J. P. Yates, the office engineer.

Only through the unified effort of these men and of countless others has it been possible for Mr. Crowe to achieve the high degree of coördination which is reflected in the noteworthy records that are being made. The completed dam not only will bear testimony to the sound judgment of those courageous individuals who envisioned the structure many years ago but will also constitute an enduring monument to the unity of purpose and unflagging energy of the personnel of Six Companies Inc., from the highest to the lowest.

**THE MILLIONTH YARD**

**Work paused long enough on January 7, 1934, to permit photographing the millionth yard of concrete placed in the forms. Those lined up in the foreground are, left to right: Charles A. Shea, director of construction, Six Companies Inc.; Felix Kahn, treasurer, Six Companies Inc.; Walker R. Young, construction engineer, Bureau of Reclamation; Bernard "Woody" Williams, assistant general superintendent, Six Companies Inc.; John C. Page, office engineer, Bureau of Reclamation; and Francis T. Crowe, general superintendent, Six Companies Inc.**

## MOTION PICTURES TELL STORY OF BOULDER DAM

THE interest in Boulder Dam throughout the country has created a surprising demand for motion pictures of the construction operations. In an effort better to meet the many requests for the official Government progression pictures, the handling of them has been turned over to a private enterprise designated the Boulder Dam Film Library. This concern is doing business in Boulder City, Nev., under a Government permit which binds it to carry a sufficient number of prints of the film to supply the calls that are received. The films, which are available in either 16 mm. or 35 mm. size, cover all phases of the work completed to date. Their 4,000 feet of length represents the edited version of some 16,000 feet that has so far been taken. A nominal rental charge is made for these pictures. Short films, 100 to 400 feet long, which depict individual phases of the operations, also are available.

This motion-picture record of the work is being produced under the direction of the Boulder Dam Film Board which is composed of representatives of the U. S. Bureau of Reclamation, Six Companies Inc., and The Babcock & Wilcox Company. It is expected that the film will be approximately 8,000 feet in length by the time the project is completed.

# Construction of the Boulder Dam*

## A Description of the Himix Concrete Plant and of the Cement Blending and Handling Equipment

LAWRENCE P. SOWLES†

HIMIX CONTROL PLANT

The concrete mixing operations are controlled from this station. Just beyond the operator is the water batcher, and in front of him is the cement batcher. On the right are the batch meter and the timing device.

THE placing of concrete in the Boulder Dam is an operation of such magnitude that it is being followed with much interest by a great many people. But behind it there are other operations which, though of a preparatory nature, are equally worthy of attention. Trainloads of gravel washed and screened to five different sizes must be delivered to the mixing plants with such regularity that there is always ample material on hand. Cement arriving by trainloads from four different mills must be unloaded and blended before it can be used. Water must be taken from the river, settled, and clarified, and pumped again to storage tanks above the mixing plants where it becomes an important constituent of the concrete. The various materials have to be mixed at the two plants built for this purpose—at the Lomix plant, which has provided all the concrete for the dam below and up to elevation 720, as well as some poured between elevation 720 and approximately 950—and at the Himix plant, which is supplying the remainder required between elevation 720 and 950 and all the concrete from elevation 950 to the top.

The Himix plant, which is now the center of these activities, will serve well to illustrate the operations preceding the delivery of concrete at the dam and also the manifold ways in which power in the form of electrical energy and compressed air is directed and coördinated so as to produce millions of cubic yards of concrete with unprecedented speed. About 5,000,000 barrels of cement, equivalent to 20,000,000 sacks, will be used in the construction of the dam and appurtenant works, and must be handled at rates as high as 10,500 barrels—about 35 carloads—per day.

The high-level or Himix plant is located at elevation approximately 1,232 on the Nevada rim of the canyon that offered the only available open space of sufficient area near the dam site to accommodate it. The plant is adjacent to the roadway which will ultimately be carried across the top of Boulder Dam and at the river terminus of the U. S. Construction Railroad. Grouped around it are a number of subsidiary structures—the plant where the cement received from four mills is blended into a material of uniform color and properties; a transport system which delivers the blended cement into the Himix silos and, through more than a mile of piping, into the Lomix storage bins; a screening plant which provides the fine cement necessary for the proper grouting of the contraction joints in the dam; and the shops required for the maintenance of these several plants and of the railroad system that hauls the concrete from Himix to the dam.

The Himix plant has been designed not only to produce concrete that will meet the rigid Government specifications set for the Boulder Dam concrete but also to deliver it at rates that will meet the contractors' schedule. Mechanically it is similar to the Lomix plant, which was previously described in an article of this series—the only

†Engineering Department, Six Companies Inc.

*Eighteenth of a series of articles on the Colorado River and the building of the Boulder (formerly Hoover) Dam.

material difference being in the arrangement and layout. In building the plant, the advantages offered by the topography have been turned to best account with the result that the concrete flows by gravity and in as nearly straight a line as possible. The aggregates are delivered by a railroad running out on a trestle built over bins so that they can be dumped directly into them from the cars. The bins are of timber construction retained on the sides by steel beams and tie members that are integral parts of the steel railroad trestle, and rest on a reinforced-concrete floor and columns. The batchers, which weigh out the various aggregates for the mix in the exact proportions determined by the Government engineers, are hung from the bottom of the concrete floor. One set of batchers supplies each mixer, and the several units are arranged in a line so that the materials drop on to a belt conveyor that delivers them to the hopper feeding their mixer.

When busy at their assigned job, these automatic batchers are fascinating to watch. The entire operation is, of course, under the control of the man at the mixer, but the apparatus is so interlocked that it is necessary for him simply to push a button to release the right quantities of aggregates from the batchers in such a sequence that the fine material falls first on to the belt conveyor and provides a cushion for the coarser material that follows. This done, the discharge doors close automatically and latch, if they are free from obstructions. This latching sets in action a solenoid valve that controls the flow of compressed air to an air-operated gate through which aggregates are again admitted into the batchers from the bins above. As the material in a batcher approaches the predetermined weight, the gate partly closes and cuts down the flow: it does not close completely, however, until the precise amount has been measured out into the batcher. The noise accompanying this operation, which is induced by the rapid opening and closing of the compressed-air gate valve and the falling of the rocks into the eight batchers, is so ear-splitting that when it suddenly ceases, and everything becomes silent, one cannot help being impressed by the fact that the compressed-air and electrical-control devices have done very exacting work at the touch of a button by an operator remote from the batchers. These, as well as the Lomix batchers, were supplied by C. S. Johnson Company, Champaign, Ill.

Four mixers are mounted upon a large block of reinforced concrete that also serves as a support for four cement silos, one above each mixer. These silos are steel cylinders, 23 feet in diameter by 48 feet high, having conical bottoms to which the cement batchers are directly connected. Each has a capacity of 5,000 barrels of cement, equivalent to 20,000 sacks.

Cement and water in the right proportions are weighed out automatically and discharged from weigh batchers into the several adjacent mixers by the operator. Behind him is a recorder and a control board. The recorder indicates the weight of each of the different materials in a batch and the hour it was fed to the mixer. When all have been delivered and mixed the specified length of time, a clock automatically signals to the operator that the batch in the mixer is ready to be turned out.

The mixers are of the 4-yard tilting type. The first two installed were Smith mixers. Later were added two Davis mixers which had been used originally for mixing concrete on the Owyhee Dam, and, finally, one Smith mixer was removed from Himix and placed between the bins and the mixer block to deliver concrete to No. 9 Cableway by means of an auxiliary track constructed behind the mixer block. Each mixer is so mounted that its whole frame tips, allowing the contents to be discharged cleanly in a few seconds. Below each is a 9-yard hopper with an air-controlled slide-gate outlet. Operating independently, a mixer can thus prepare two 4-yard batches of concrete and load them into its hopper ready to fill without delay one of the 8-yard concrete buckets when it reaches the mixing plant on its return from the dam. In this manner little time is required for loading at the plant. This has proved to be a decidedly important factor in maintaining the high concrete-production records so far made.

Cement is brought to the project in standard 40-foot box cars especially selected for the service and fitted with special floors and spouts in the roofs that permit filling the cars at the mills by gravity. The cars are run on to a siding adjacent to the

UNLOADING CEMENT PNEUMATICALLY

Box cars filled with cement come in on a siding above the Himix plant. Between the two tracks is a platform, which is shown above. By means of Fuller-Kinyon air-operated pumps the cement is transferred at the rate of 150 barrels an hour into blending silos. These pumps are mounted on wheels to facilitate moving them as required. At the left is one of these units in a car of cement.

CEMENT TRANSFER LINE

The 9-inch pipe line through which blended cement is delivered from the Himix to the Lomix plant traverses rough ground in reaching the canyon bottom (right). At the upper left in the picture is the 180-foot boom of the derrick that handles concrete for the Nevada intake tower. Above is the lower terminus of the cement line at the Lomix plant.

Himix plant where the bulk cement is unloaded by a Fuller-Kinyon pneumatic transport system. The cars have timber partitions so set that there is an open space between the two doors. This enables removing the bulkheads and starting the unloading machines without cement wastage.

The siding provided for this purpose comprises two parallel tracks, approximately six cars in length, and a covered 13-foot timber platform between them. The unloading system consists essentially of Fuller-Kinyon cement pumps each of which is mounted on wheels in such a way that it can be directed by an operator into the pile before him. The cement discharged from a pump is conveyed through a 5-inch line to one of the silos at the blending plant. Three pumps handle the normal demand, with two pumps held in reserve as spares. One machine unloads a 300-barrel car in about two hours. Normal cement consumption necessitates the unloading of about 30 cars a day, working three shifts. The fundamental principle upon which this system operates is that cement—finely ground powder—when mixed with a sufficient amount of air, becomes a semifluid material and flows much like water or any other fluid. The compressed air provides the force that drives the cement through the conveyor and into the silo.

The 5-inch lines from the three cement unloaders are carried across to the blending plant and are manifolded so that any one pump can deliver into any one of the eight silos. When the cement arrives at the bins, the velocity of the mixture drops sufficiently to allow all but a minute amount of the cement to settle out of the air. Attached to each bin is a vent which, as the cement is deposited in the bin, permits exhausting the air that was introduced into the pump when fluffing the cement. A manhole in the roof of each bin provides access for cleaning, and it also enables the operator to check the depth of the contained cement when filling a silo.

The blending plant consists, in the main, of the cement unloading system, of the eight raw-cement silos, and of a set of variable-speed feeders. The latter control the discharge from the various silos into a long screw conveyor which mixes the cement as it carries it to the bin of the cement transport system. Each silo is $21\frac{1}{2}$ feet in diameter and 48 feet high, and holds approximately 6,000 barrels of cement. The correct rate of feeding each brand of cement from the blending silos is determined by Government inspectors, and the result of their calculations is given to the operator of the plant who sets the variable-speed drive on each rotary feeder accordingly. The feeders have a range of from 50 to 200 barrels per hour.

During the study of the problem of handling and blending the cement it became evident that, from the standpoint of economy, the plant for this operation

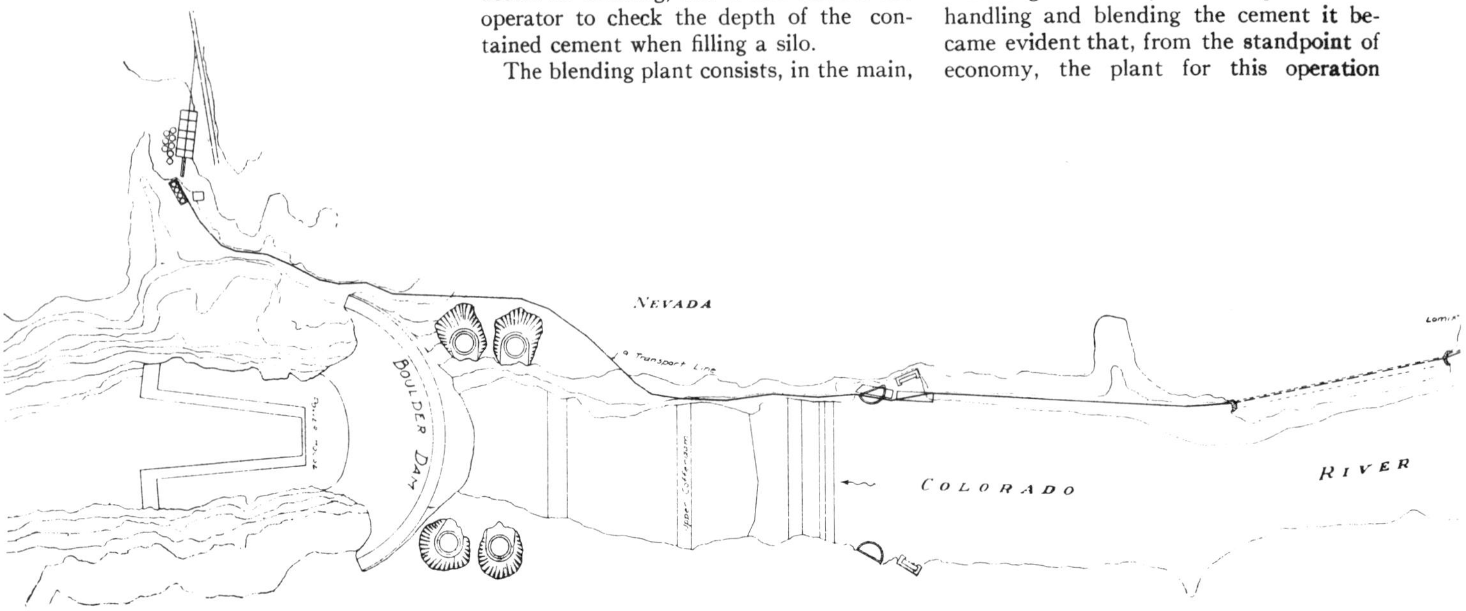

### SECOND-STAGE FLUXO PUMP

As installed, the cement pumping system has a booster Fluxo pump in the canyon bottom at a point 2,460 feet from the primary pump. The view at the left shows the booster station with the housing designed to collect the cement from the air upon being discharged from the incoming pipe line. The direction of movement is towards the right. After the system had been operating for six months it was found that the pump at Himix would deliver cement at a slightly reduced rate to the Lomix plant. As a result this booster station is now used only intermittently. At the right is a close view of the No. 2 or booster Fluxo with its control box. The No. 1 unit at Himix is a duplicate of this one.

should be located adjacent to one of the two mixing plants. However, this meant transporting cement from there to the other plant. Upon investigation it was found that this could be done at a cost that would not be prohibitive by the use of a Fluxo pump system. By this system the cement from the Himix blending plant is delivered to the Lomix plant in either one or two stages through a 9-inch line. The first-stage pump is located at the Himix mixer block, and the second-stage pump is situated on top of the upper cofferdam on the Nevada side of the canyon.

In laying out the Fluxo system the 9-inch transport line had to be carried over the difficult topography along the high-level road, across several deep draws on cable spans, and down the side of the Nevada Canyon wall to the bin for the second-stage pump. From there it had to be run along the canyon wall, across the top of the double-track bridge on the concreting railroad at elevation 720, and then along the track and through No. 1 Railroad Tunnel to the Lomix plant. From the outlet of the tunnel to the Lomix plant the line is supported upon a cable span approximately 300 feet long and rising about 105 feet to the top of the cement bin. Interposed at this point in the line is a valve which permits emptying the cement into the bin at either one of two levels. The total pumping distance is 5,420 feet—the first stage being 2,460 feet long and the second 2,960. After a working period of approximately six months, it seemed practicable, by slightly cutting down the hourly capacity, to pump directly from the blending plant to the Lomix plant in a single stage. This cut-over was tried, and was very successful.

In the operation of the Fluxo pump, we have an unusual application of compressed air in the field of materials handling. Pumping is accomplished entirely by the energy exerted by the air, which serves also to fluff and to make the cement semifluid. The pump, itself, consists of two cone-shaped, bottom-pressure tanks 6 feet in diameter and 9 feet high. These are interconnected electrically and mechanically so that one is discharging while the other is filling. The various interlocks regulate the time of admission and cut-off of the compressed air to the tanks. The system is operated by means of a series of relays containing a number of Mercoid electrical switches which, in turn, actuate solenoid valves that control the air supply to and from the tanks. The pumps were manufactured by the F. L. Smidth Company, of New York City, and installed by Six

### CEMENT TRANSPORT SYSTEM

At the left is the route of the 9-inch pipe line through which cement is transferred from the Himix to the Lomix plant by means of a Fluxo air-operated pumping system. During the course of its movement, the cement travels 5,420 feet and drops more than 500 feet. This installation makes it possible to blend all the cement at the upper plant and eliminates a circuitous rail haul of some fifteen miles into the canyon. The layout of the Himix plant is shown at the right. During the course of the work, 20,000,000 sacks of cement will be blended there.

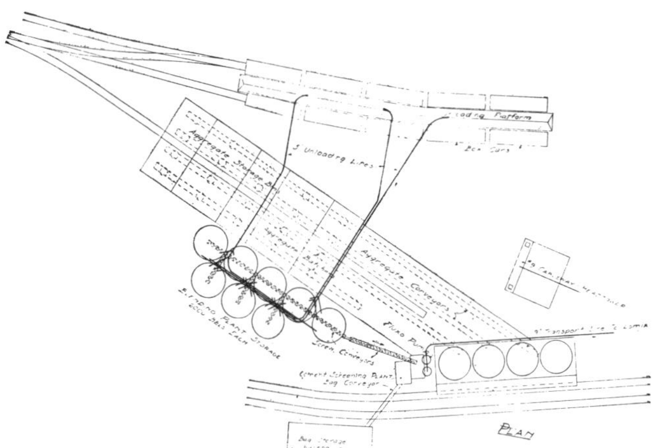

Companies Inc. under the direction of Smidth Company engineers. Each transports 450 barrels of cement an hour. The first-stage Fluxo pump, in addition to delivering cement to the Lomix plant, also delivers it, through a 6-inch line, to any of the four silos at the Himix plant, or, through another 6-inch conveyor, to the bin of Himix "E" mixer. The line first referred to is manifolded on top of the silos so as to reach any one of them.

As a result of the tests made at the Denver office of the U. S. Bureau of Reclamation, it was found desirable to employ a very finely ground cement for the grout for the contraction joints in the dam because it would assure a more uniform filling of the joints, a more uniform thickness of grout in each joint, and a better penetration of the grout to every part of a joint. But instead of specifying the use of a finely ground cement that would pass the requirements, it was decided to take the blended cement that goes into the concrete for the dam and screen it, thus removing from it all material that will not pass a No. 200 screen.

Six Companies Inc. have constructed a screening plant for this purpose next to the No. 1 Fluxo bin. Blended cement is drawn from the nearest Himix silo through a feeder into the primary screen— a 6x5-foot Tyler Hum-mer. The rejects from this are fed to another Tyler Hum-mer screen that measures 3x5 feet. The product which passes these screens goes into a bin from which it is sacked in paper bags to simplify handling the cement in its transfer to points of use. When a bag has been filled by the packer, it is delivered to a slat conveyor which carries it to a screened storage shed a short distance from the plant. This shed holds 2,000 sacks. The plant has a rated capacity of 2,600 pounds of cement per hour, and the present estimated requirement for contraction-joint grouting is 26,000 sacks. The Bureau of Reclamation has specified that screened cement for grouting shall not be stored longer than seven days after screening or before use. A plant of the aforementioned capacity is therefore needed in order that all the cement for any one grouting lift can be produced within the 7-day limit set.

The plants described were designed and laid out by the Engineering Department of Six Companies Inc. under the supervision of A. H. Ayers, chief engineer, and J. P. Yates, office engineer, and, with the exception of the patented equipment, were constructed by that company under the direction of F. T. Crowe, general superintendent, and Bernard Williams, assistant general superintendent. Ira Carpenter, concrete production foreman, is in charge of the operation and maintenance of the Lomix and Himix plants, as well as of the concrete-blending, transporting, and screening plants.

GENERAL VIEW OF THE HIMIX PLANT

This photograph shows the installation as it now stands. The four silos to the right each hold 20,000 sacks of cement. Below them is the mixing plant, containing four 4-cubic-yard units. One of them is shown filling an 8-cubic-yard bucket stationed on a car. The chute across the tracks at the lower right is for loading agitators. The silos at the upper left are for blending cement. Between them and the mixing plant proper may be seen a fifth concrete mixer which supplies cars on a track leading to No. 9 Cableway. At the top are cars on the cement unloading siding. The cables crossing the picture are backstays for the Government 150-ton cableway.

FIRST TRIP OF TRAILER

Delivering the first section of 30-foot-diameter penstock header pipe from the Babcock & Wilcox fabricating plant to the rim of the canyon. The trailer and its load weigh more than 200 tons.

# Huge Trailer Hauls Penstock Pipe at Boulder Dam*

COPELAND LAKE

BECAUSE Boulder Dam is the largest structure of its kind ever built, it logically follows that its construction calls for doing various essential operations on an unprecedented scale and with equipment of unusual size. Descriptions of the work accordingly involve the frequent use of superlatives. In previous articles of this series attention has been directed to the 56-foot diversion tunnels and to the huge 30-drill "jumbos" that played such an important part in their excavation. We have likewise told about the record-size cableway system, the facilities for preparing concrete aggregates faster and in greater quantities than was ever done before on a construction contract, and have described the equipment and methods by which concrete is being mixed and placed in the forms at a rate never before attained.

Similarly, as we follow the story of this project through to its conclusion, we shall find other elements and features that surpass in magnitude all others of their respective kinds that have gone before. Notable among them may be mentioned the hydraulic turbine-generators and the steel conduits which will carry water to them. The present article has to do with the latter structures in an incidental way, inasmuch as it relates how the enormous erection sections of steel pipe are moved to the rim of Black Canyon preparatory to being lowered into position in the rock tunnels that penetrate the cliffs. Here, again, we meet equipment that outstrips anything of the nature previously built.

The fabrication of the steel pipe for the penstock headers and penstocks and its installation within the tunnels constitute an undertaking of such size that the Bureau of Reclamation elected to make it a separate operation and to call for bids covering the work. The contract, which is for approximately $11,000,000, was awarded The Babcock & Wilcox Company of Barberton, Ohio, and that well-known and experienced manufacturer of steam boilers and other types of pressure vessels has been engaged in carrying it out for the past two years. Its task consists of forming steel conduits aggregating 14,500 feet in length and of assembling them within the rocky walls of the canyon. Depending upon the service for which they are intended, these conduits will be of four sizes, their respective diameters being 8½, 13, 25, and 30 feet. They will have wall thicknesses ranging from ⅝ inch to ⅞ inch for the smallest size to from 1$\frac{11}{16}$ inches to 2¾

*Nineteenth of a series of articles on the Colorado River and the building of the Boulder (formerly Hoover) Dam.

inches for the 30-foot size. All these are being made up of flat plates and will require the use of 45,000 tons of material.

Owing to the fact that the larger sizes of pipe are too big, even when built up in small sections, to transport by railroad, it became necessary for The Babcock & Wilcox Company to erect and equip a large fabricating plant near the dam site. The closest suitable location available was at Bechtel, Nev., about a mile from the canyon rim. There, in desert surroundings, is being done work of a sort that is usually associated only with industrial areas.

One of the accessory but vitally important features that had to be considered in connection with this work was the devising of adequate methods and facilities for moving the huge pipe sections from the fabricating plant to the edge of the canyon for delivery to the 150-ton cableway which stood ready to lower them to any one of four shelves leading to various parts of the tunnel system within the cliffs. Inasmuch as Six Companies Inc. was experienced in problems of this sort and had an organization on the ground, arrangements were made whereby it would take over this transportation task.

The true proportions of the difficulties involved can be comprehended by considering the form and characteristics of an erection section of a 30-foot conduit. A unit of this type consists of two rings, each 30 feet in diameter and 12 feet long, joined end to end and reinforced at their joint with a stiffening member. Each of these two component rings is made of three plates each 12 feet wide, approximately 31½ feet long, and, in the case of the heaviest pipe, 2¾ inches thick. Such plates weigh 23 tons apiece, and when six of them are knitted together the resulting erection section weighs about 170 tons.

There were only two ways by which these sections could possibly be moved—by railroad and by highway. In the case of the former, it was determined that four or six rails would be required to secure lateral stability, involving the laying of additional trackage and the widening of several deep rock cuts on the single-track railroad. Inasmuch as the existing roadway could be used with no changes other than a reduction in the super elevation on some of the curves, the decision was made in favor of the highway, and steps were taken to secure a vehicle of suitable size, strength, and maneuverability to carry the large and heavy loads. The conditions pointed to a trailer as being the most acceptable form of carrier, and no time was lost in having one made to meet the needs.

The trailer was furnished by the C. R. Jahn Company, of Chicago, Ill., and was planned and built by the La Crosse Boiler Company of La Crosse, Wis. It has a framework of steel and is 37 feet 8 inches long by 22 feet wide. It has four wheels at each corner, and is equipped with a hydraulic steering mechanism and with air-operated brakes. Without load, the trailer weighs 41 tons It was designed to carry a maximum load of 200 tons.

The road from the fabricating plant to the cableway is 1½ miles long and is downhill all the way towards the canyon with a maximum grade of 6½ per cent. Its overall width is 30 feet, of which about 28 feet is surfaced with oil-treated gravel. It has numerous curves, one of which has a radius of 100 feet on the center line, and a grade of 4 per cent. Because of a lack of space in which to turn so large a trailer around, and also because it was recognized that turning it would be attended by certain difficulties and dangers, the unit was made reversible and was equipped with drawbars for hauling from either end. Tractors are employed for the towing.

**BEGINNING 600-FOOT DESCENT**

**The first section of 30-foot pipe about to go over the wall for lowering into the canyon by the Government 150-ton cableway. The weight is suspended from special "moonbeam" lifting devices designed to equalize the load upon the track carriage. The holes which are visible in the pipe wall will be used for filling the space around the conduit after it has been set in place with concrete.**

As the length of an erection section of 30-foot-diameter pipe is 24 feet, and the center of gravity is quite high, wide spacing of the wheels was called for and this, in turn, produced problems in connection with steering the trailer around curves. It was obvious that conventional methods of steering were inadequate under such conditions, because in turning a curve with a 100-foot radius the wheels on the outer arc were certain to be dragged to some extent in keeping up with those on the inner or short arc. This, of course, could not be allowed to happen under such a heavy load because of the disastrous effect it would have on the tires.

As can be noted in one of the accompanying pictures, the four wheels at each corner are supported on two axles. These axles are of the tilting, oscillating type which balances the load and compensates for ordinary road irregularities without materially changing the load-carrying level. The wheels are mounted on Timken double, tapered roller bearings which are lubricated by a high-pressure grease system. On each wheel are two 28x14-inch solid Goodyear tires. These provide a total of 448 inches of tire space on the sixteen wheels. The carrying capacity of each tire is 10,200 pounds, with an overload allowance of 25 per cent in view of the slow speed at which the trailer travels.

To facilitate turning, each set of dual axles is independently controlled by the steering apparatus. A Northern nitro-alloy steel pump, driven by a 15-hp. gasoline engine and having a capacity of 50 gpm. of oil against 300 pounds pressure, supplies constant equalized pressure to a control cylinder at each set of axles. By means of a compensating link between the two main steering levers, it is possible when on curves to adjust the radius of travel of the inner wheels so as to give all sixteen wheels a rolling action. Steering is controlled through a wheel mounted at one end of the trailer.

As the road is down grade practically all the way from the fabricating plant to the canyon rim, the use of brakes is well-nigh continuous when the heavily loaded trailer is moving in that direction. This naturally makes it imperative that the compressor and other elements in the braking system be of unfailing reliability and efficiency. Bendix-Westinghouse internal-expanding-type brake chambers are mounted on each wheel. They are supplied with compressed air by an Ingersoll-Rand Type 30, air-cooled compressor which is driven by a 5-hp. gasoline engine. This unit has a maximum discharge pressure of 200 pounds per square inch and maintains a constant pressure of from 90 to 100 pounds. In addition to the main air receiver located near the compressor, auxiliary storage tanks are placed at each corner of the trailer. The total braking surface is 3,166 square inches. Emergency features permit stopping the trailer instantly should the towing unit become disconnected or a break occur in any of the air lines to the wheels.

Rubber-covered cradles provide resting places for the pipe—various sizes of these supports being available to accommodate the several different-sized sections.

# Construction of the Boulder Dam[*]

## Government Engineers and Surveyors Have Made a Notable Record on Exacting, Perilous Work

WESLEY R. NELSON[†]

**ALPINEERING ON THE ANDESITE**

**A survey party on the steep cliff of the Nevada abutment while excavating in that area was in progress. A bos'n's chair used by one of the scalers working at this location appears at the left.**

ADVANCING in the forefront of progress, taking the brunt of pioneering hardships, and at many times securing necessary data even at life's peril, will be found the explorer. the engineer, and the surveyor. Particularly is this true in the studies, examinations, and surveys conducted along the Colorado River, that turbulent stream which has drowned scores in its rapids and well earned its appellation of "The most dangerous river."

As Dr. Elwood Mead, Commissioner of the U. S. Bureau of Reclamation, has written, "The search for a suitable site for Boulder Dam went on for many years. Its history is filled with dramatic incidents. The early explorations were fraught with unknown dangers. Lives were lost, and there was always present a grave possibility that those who dared to traverse the gloomy canyons would never live to report their findings. Lieut. Joseph C. Ives of the U. S. Army, exploring the river in 1857, gave up with the conclusion that the Colorado, along the greater part of its way, should be forever unvisited and undisturbed.

"The survey of the dam site and reservoir was of unprecedented magnitude and difficulty. It involved coping with a river which, in the highest floods, rushed through the canyon with the speed of a railway train, and taking topography in more than

---

*Twentieth of a series of articles on the Colorado River and the building of the Boulder (formerly Hoover) Dam.

†Assistant Engineer, U. S. Bureau of Reclamation, Boulder Canyon Project.

ALL IN THE DAY'S WORK

Suspended from the canyon rim, a rodman or "rigger" indicates with a 15-foot pole a control point for transit shots from the opposite cliff. During the making of a topographic survey in 1932, rodmen in this manner descended cliffs as much as 1,000 feet high. Where there were overhangs or caves, these men sometimes had to swing themselves inward with a pendulumlike motion to furnish readings at recessed points. This survey party consisted of a chief, two transitmen, two rodmen, two recorders, two ropemen to lower the rodmen, one man in the canyon to warn workmen of falling dislodged rocks, and one signalman stationed between the two transits to establish communication between the transitmen and rodmen.

100 miles of canyon where precipitous cliffs 1,000 feet high and of indescribable ruggedness had to be scaled. Three lives were lost in this hazardous undertaking. Every phase of the work involved great danger, but the dimensions of the possible dam and reservoir had to be known. Then there had to be a topographic map.

"Anyone who views the canyon either from the top of the rim or from the river at the bottom, has a sense of the peril and hardship involved in fixing locations and making measurements on its cliffs. To have done this work by the old methods would have delayed beginning construction six months to a year. Resort was had to aerial surveys. This involved great hardship because of the intense summer heat and making observations at great differences in elevation."

Investigations by the United States Government of the potential resources of the Colorado have been in progress since 1850, when the sailing vessel *Invincible*, under the command of Lieut. George B. Derby, U. S. Army, penetrated the river from the Gulf of California to the mouth of the Gila River. Seven years later, Lieutenant Ives assembled a stern-wheeler, the *Explorer*, at the mouth of the Colorado and traveled on its waters to the upper end of Black Canyon where a sunken rock damaged the boat and prevented further progress upstream.

Maj. John Wesley Powell outfitted an expedition at Green River, Utah, in 1868-69, and descended the Green and Colorado rivers by boat to the mouth of the Virgin River, 35 miles upstream from Black Canyon. In 1870 Major Powell was placed in charge of an organization to make topographical and geological surveys along the Colorado and its tributaries. The survey, when completed, covered more than 100,000 square miles of territory.

The Department of the Interior, through the Bureau of Reclamation and the Geological Survey, has been making investigations and surveys on the Colorado River for the past 31 years, determining its characteristics, analyzing its resources with reference to the development of lands in its basin, and locating sites where the construction of dams is feasible from structural, geological, and economic standpoints. As a result of this work, irrigating water has been provided or supplemented for hundreds of thousands of acres of arable land, thousands of homes have been established, and millions of dollars have been added to the national wealth.

Notable among the irrigation projects that have been undertaken are the Grand Valley and Uncompahgre in Colorado; the Strawberry Valley in Utah; the Salt River in Arizona; and the Yuma in Arizona and California. Only one of these, the Yuma, secures water from the main Colorado in the lower basin. Several other irrigation projects have been developed in that section through private endeavor; but the inadequate flow of the river in some years, the heavy silt load it carries, and the destructive tendencies when in flood, have restricted the settlement of lands, burdened the promoters with costs amounting to millions of dollars for cleaning deposited silt out of canals, laterals, and ditches, and threatened complete inundation or destruction with a resultant loss of confidence and of credit.

With these facts in mind, special consideration was given to a dam site in the lower basin that would provide a large reservoir for the storage of water and silt, that would effectively protect the downstream areas from flood, and that could be constructed without burdening the benefited lands with a debt that could not be paid under existing agricultural conditions.

Parties of Government and private engineers that had traveled down the tortuous river had made mention of feasible dam sites at Boulder and Black canyons as far back as 1902. It was, however, not until 1919 that the late Arthur P. Davis, then director of the Reclamation Service, envisioned the possibility of power as an aid to the enterprise, and he courageously proceeded to plan a dam of unprecedented height and a reservoir of unprecedented capacity. Engineer-Geologist Homer Hamlin, in charge of a party sent out in 1920, placed a stake to mark the place in Black Canyon recommended by the director for the building of a dam. That stake has since been found within a few feet of the axis of the dam now under construction.

In 1919, the U. S. Geological Survey was directed by the Secretary of the Interior to make a topographical survey of the proposed reservoir site upstream from Boulder Canyon, and to take elevations for the plotting of contours at 50-foot intervals from the river up to elevation 1,250. From such a survey could be calculated not only the volume of water backed up by dams with varying crest elevations but also the area of land that would be inundated.

The party that was organized for this purpose was compelled to use the river as a means of transportation, as the canyon-wall terrain was far too rough and precipitous for travel by land. The survey proved to be of a kind that taxed the endurance of all its members, and presented grave hazards and risks each day the work was carried forward.

The Colorado is notorious for its dangerous rapids that have blocked successful navigation of that stream since the first attempts were made 400 years ago. Torrential rains sweep bowlders from side canyons into and across the river channel, forming rock barriers over which the water breaks in a thousand cataracts or is turned in a mad frenzy toward the canyon walls, rushing by at the speed of a mill race. When large bowlders lie just beneath the surface, "holes" are formed on their downstream sides into which the water drops and churns in a roaring crescendo.

The party was able to portage past many of the rapids and to line the boats through at points where talus slopes jutted into the stream at the base of cliffs; but in many places the walls rose abruptly from the water's edge and the only route was through the rapids. The boats used on the expedition were generally about 25 feet in length and had hatches forward, amidships, and aft for supplies and equipment which were sheltered by tight-fitting hatch covers. When shooting rapids, the stern of the vessel was turned downstream, guided into the deepest water at the head of the rapids, and then relinquished to the current as the boatman fended off projecting rocks, dodged "holes" and whirlpools, kept the craft on an even keel, and headed downstream in a swift descent through the turbulent swirling waters and rock-strewn channel. A partial boatload of water was usually shipped in the passage, and boats were often overturned. Other dangers lurked in the swift rising floods; and the men had to be continually on the alert when encamped on the river shore because rains in the lower elevations or melting snows in the mountains were apt, during any month in the year, to precipitate a torrent, causing the water's surface to rise as much as a foot in an hour and doubling the flow in a day's time. Added to the ever-present perils of the river were the disagreeable weather

CLOSE CHECKING

**The stake shown was placed in 1920 by a survey party in charge of Engineer-Geologist Homer Hamlin to mark the location recommended for the construction of a dam in Black Canyon. It was found years later by engineers who fixed the axis of the Boulder Dam at the point indicated with white paint above and to the left of the stake.**

conditions that existed in the desert region. The cold blasts of winter were followed by winds of high velocity in the spring, and these were succeeded by the furnacelike heat of the summer when thermometers registered as high as 128°F. in the shade.

As a result of recommendations made by the Bureau of Reclamation, based on these investigations and surveys, Congress in 1920 passed the Kincaid Act authorizing and instructing the Secretary of the Interior to make examinations and to report on the problems of the lower Colorado River.

Explorations of foundations by diamond-drilling were started in Boulder Canyon soon after under the direction of the superintendent of the Yuma Project. On January, 1921, they were taken in charge by Engineer Walker R. Young, who continued the work until May of that year when he was forced to suspend operations owing to high water. These investigations were resumed in the winter and spring of 1921-22, and completed in the same seasons of 1922-23.

With the exception of angle holes and a few other holes along the shore line, all the drilling was done from derricks mounted on barges. Each barge was constructed of two pontoons, 36 feet long, on which was laid a deck 15 feet wide. The barge was hung from cables spanning the waterway, and was moved to the shore when the river rose in flood. The drills, equipped with diamond bits, were rotated by 10-hp. gasoline en-

TEN YEARS BEFORE CONSTRUCTION BEGAN

Three barges equipped with diamond-drill rigs testing the rock beneath the river at Boulder Canyon in February, 1921. Investigations of possible dam sites went on for many years before the Black Canyon location was selected. The capriciousness for which the Colorado is noted rendered these activities highly dangerous at times.

gines. Holes were usually drilled on lines 200 feet apart and to depths as great as 200 feet extending 50 feet into solid rock.

The personnel at Boulder Canyon during the first drilling season numbered as many as 58 men; the camp consisted of 28 tents; and four barges were employed in the drilling operations. A group of flat-bottomed boats, several equipped with outboard motors, carried the men and supplies to the barges.

Drilling in the Colorado was never expected to be a sinecure: but it soon became evident that the river's power and strength had to be treated with respect, and that slight miscalculations might result disastrously, even have fatal consequences. Quoting from Engineer Young's report for February, 1921, to the Denver office: "Two men in a small boat were swept in front of and under barge No. 2 on the 22nd. Men and boat were recovered without loss. Man fell off of barge No. 1 on the 23rd, but caught a rope. As the night shift was leaving the C line for camp on the 28th their motor stopped working and the boat with four men was swept in front of and under barge No. 3. Three of the men clung to the barge, but the fourth went under. He was given up as lost, but was picked up about three hours later on the Arizona side of the Canyon by a boat sent down the river from camp. The boat and motor were ost."

At another time a barge broke loose from its cable, throwing three men overboard. Driftwood, carried so plentifully by the river, broke standpipes beneath the barges, or caused the latter to sink with a loss of all their equipment. Storms occurring in December, 1921, washed out road connections between the camp and St. Thomas, the nearest town where supplies were obtained; and the rise in the river resulting from the flood destroyed one barge and drowned one of the drillers.

In January, 1922, a camp was erected near the mouth of Black Canyon and investigations were started of the foundations at the upper dam site. A windstorm that arose soon after leveled the camp, blowing away most of the equipment even down to the floor boards of the tents. The camp was rebuilt and work continued until May. It was then suspended for the summer, resumed in September, and completed on April 24, 1923.

At the time drilling was in progress, topographical surveys were being conducted in the river channel and on the sheer 1,000-foot-high canyon walls to secure data for the plotting of maps and the subsequent designing and locating of the proposed dam and related features. Trails were cut and blasted along the cliffs for this work, and ladders were erected to reach otherwise inaccessible places. During those surveys, triangulation points were established and tied into preceding land surveys. Mean sea level was used as the datum for elevations, starting from bench marks set on a line of levels run by the Geological Survey from Hackberry, Ariz., to Moapa, Nev., with a loop line extending to St. Thomas. Supplementing the surveys, geological investigations were made of the reservoir area and of the Boulder and Black canyon dam sites, suitable gravel deposits were located, and the region around each dam site studied with reference to details of proposed construction, accessibility, and related features. All data were assembled by the Bureau of Reclamation; and in 1924 a report was submitted to Congress recommending the building of a concrete arch-gravity-type dam in Black Canyon.

The principal reasons for preferring Black Canyon to Boulder Canyon were: greater accessibility, lesser maximum distance to bedrock, larger reservoir area, lesser amount of concrete required for the same height of dam, and better geological conditions particularly in relation to faults.

Then followed a period of several years when little work was in progress in or near Black Canyon, as the project was under discussion in Congress and legislative matters and irrigation problems were being straightened out in the states of the Colorado River basin. After these matters had been settled, the Boulder Canyon Project Act was passed by Congress; power contracts were negotiated to insure repayment of construction charges in 50 years; and the first appropriations were made on July 3, 1930.

Engineers of the Bureau of Reclamation and consulting boards had weighed all available information and determined that Black Canyon was the more feasible site for the construction of the dam. The preliminary surveys of that site, however, had extended only far enough to provide data for the design of a dam. As further information about the region for several miles around the dam site and more details

BLACK CANYON
This view of the dam site, guarded by forbidding ramparts of lava, conveys a fair idea of the difficulties that confronted the preliminary engineering parties.

REAL BOULDER DAM PIONEERS
Personnel of the Boulder Canyon survey camp in February, 1921. This picture was made by the chief of the party, Walker R. Young, whose photograph appears below. Mr. Young has been intimately connected with the Boulder Canyon Project for thirteen years, and is seeing the job through as construction engineer for the Bureau of Reclamation.

regarding the canyon walls were desired with a minimum of delay, additional topographic surveys were started by airplane and from the ground within two days after the appropriations had been made. These were completed within two months. The surveys by airplane were contracted for, and covered an area of 96 square miles around the dam site: those by ground methods covered approximately 8,000,000 square feet, as projected on a vertical plane, of the precipitous canyon walls. A mosaic and topographic maps were prepared from this work, using special equipment based on the principles of stereoscopic scrutiny.

As construction got underway on the project, the demand for surveys became greater and greater. From one survey crew the number grew to five, ten, and then to fourteen, the personnel from 5 to 60. The work was extremely fatiguing, as well as hazardous. Ladders were built to reach some locations, but others were found inaccessible except by lowering men with ropes. To ascend or descend the canyon walls by ordinary means usually required long hikes over trails that were few and far between.

Springtime winds, causing discomfort and danger by carrying sharp sand and by dislodging pebbles and rocks from the canyon walls, were followed by the intense heat of the summer. During July the mean temperature in the shade was 107.4°F. On two days the maximum temperature was 128°, and for two-thirds of the month the thermometer at some time during the

UNDERGROUND WORK

**A drill carriage in the Arizona penstock header tunnel, showing surveyors measuring the excavated section with "wheel" and tape. At the height of tunneling activities, Government engineers had to make measurements and supply data at upwards of fifteen headings. In the larger bores, guide marks for the next round were painted in white on the drilling face after each blast.**

day registered above 120°. Thermometers broke at 140°, rocks and metal burned the hands, and the canyon walls reflecting the sun's rays created an inferno in the depths below. All measurements required temperature corrections for chaining, surveying instruments had to be shaded, and heat waves made reading of record or point impossible for any except short distances. Surveying was started at 2 a.m. and stopped at 11 a.m. The men worked with little clothing—only shoes, trousers, and helmets. Reflecting the loyalty of the men to their organization and indicating the careful manner in which the survey was performed, it is of interest to note that none quit work on account of its arduous character, none was discharged, and none was killed or seriously injured once construction was in progress.

Great ingenuity and initiative were required in locating the numerous structures at the dam site, in measuring the quantities that were to furnish the basis for monthly payments to the contractors, and in making the check surveys after construction was finished. Tunnel headings were painted after each round of blasting, thus outlining the boundaries for the following shot. Excavation lines were painted on the canyon walls at the sites of the intake towers, spillways, valve houses, cofferdams, and the main dam structure. These were remarked when destroyed by blasts; and as an excavation approached final lines, the rocks remaining in that section were outlined for trimming operations. In some of the tunnels the usual underground surveying obstacles were accentuated by the fact that curves connecting the center lines of intersecting tunnels were neither in a vertical nor a horizontal plane. Excavations were outlined in such cases by 3-dimension graphs prepared in the field office.

The phototopographic data were not sufficiently accurate nor detailed for all purposes, hence, in 1932, a survey of the walls of Black Canyon was instituted so as to secure information for the accurate plotting of a topographic map to a scale of 20 feet to the inch.

An unusual procedure was adopted to obtain the desired data. Two transits were set up on one canyon wall at points of known location, elevation, and back sight. The rodman, equipped with a 15-foot flagpole, was lowered by rope down the opposite wall, in places nearly 1,000 feet high, and stopped at intervals to enable him to indicate with the pole the points of control for the topographic data. After reaching the canyon floor, he returned to the rim by trail or by climbing a rope, while another rodman was lowered at an adjacent location. Because of overhanging cliffs, the rodmen were required, in many instances, to swing themselves inward in pendulumlike manner in order to reach otherwise inaccessible points. As an indication of the precipitous nature of the canyon and of the exactitude of the survey, 4,000 transit shots were taken in a horizontal area of 320x660 feet, the longer dimension extending along the canyon and the other dimension toward the river. The maximum difference in elevation in this section amounted to 600 feet.

The surveyor accomplishes his important task in a quiet and unobtrusive manner, and the casual observer seldom realizes that all the construction operations are based on his maps. When watching the high scalers at work on the canyon walls, the tunnels being driven through hard rock toward definite objectives, or the concrete being poured in channel linings, intricate structures or dam columns, it is well to remember that the surveyors were there before the work was started, are there to locate and to guide construction, and will follow the project through to its conclusion.

The Bureau of Reclamation, which has so thoroughly investigated the characteristics and potential resources of the Colorado River, has as its administrative head Dr. Elwood Mead, Commissioner, at Washington, D. C. All engineering and construction is supervised from the Denver, Colo., office, where R. F. Walter is chief engineer, S. O. Harper is assistant chief engineer, and J. L. Savage is chief designing engineer. Walker R. Young, construction engineer at Boulder City, Nev., is in direct charge of field operations at Boulder Dam.

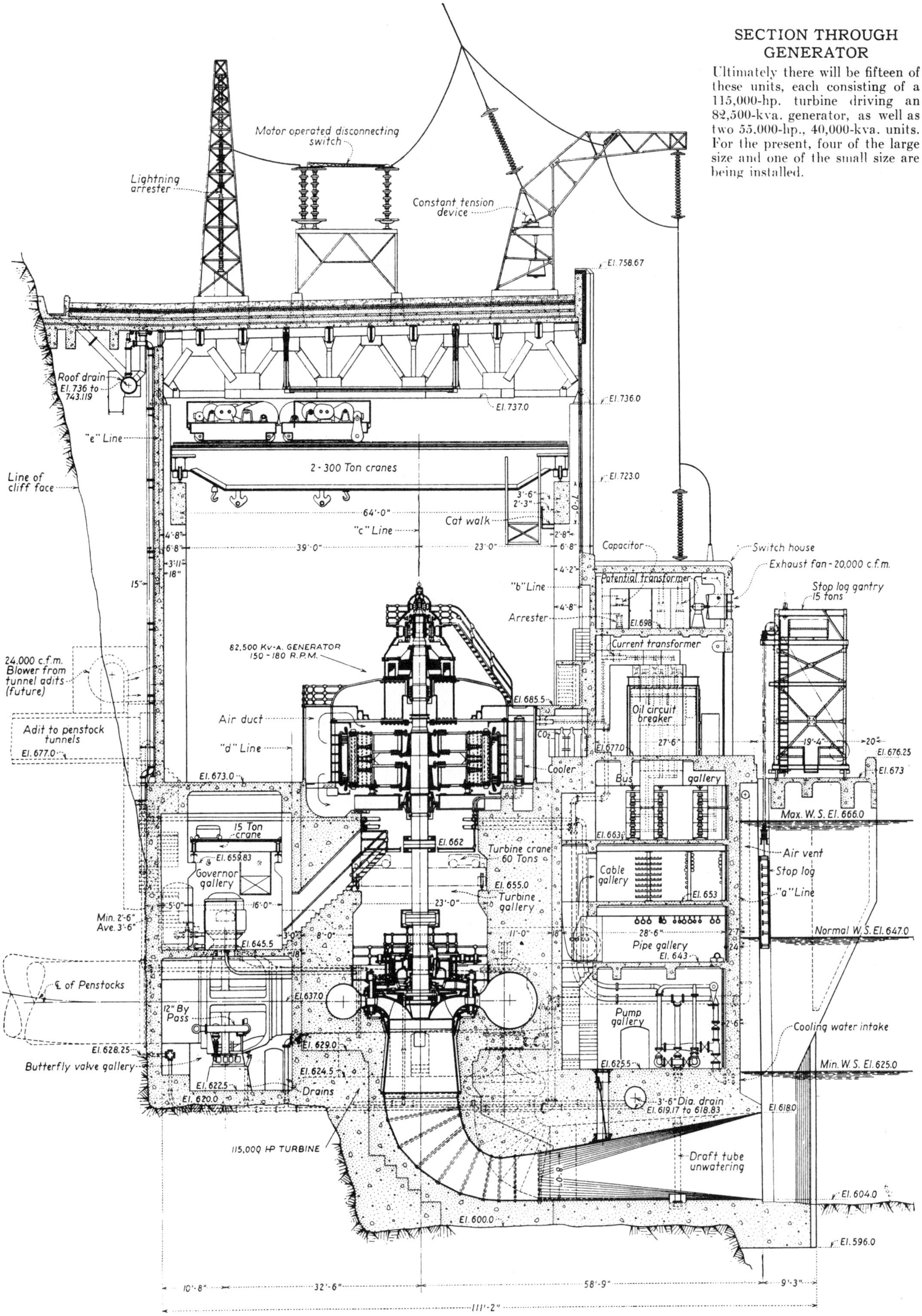


## SECTION THROUGH GENERATOR

Ultimately there will be fifteen of these units, each consisting of a 115,000-hp. turbine driving an 82,500-kva. generator, as well as two 55,000-hp., 40,000-kva. units. For the present, four of the large size and one of the small size are being installed.

# Construction of the Boulder Dam*

### How the $35,000,000 Power Plant Will Appear When Completely Equipped

Wesley R. Nelson†

*All illustrations were furnished by the U. S. Bureau of Reclamation.*

INTAKE TOWERS

Through cylindrical gates in these 390-foot-high reinforced-concrete structures, two of which are located on each side of the river, water will be taken from the reservoir and directed through steel-pipe-lined tunnels to the turbines. Each tower has an average diameter of 75 feet. When this picture was taken, July, 1935, the reservoir surface was at elevation 884, or 10 feet below the base level of the towers which are built upon shelves in the canyon walls.

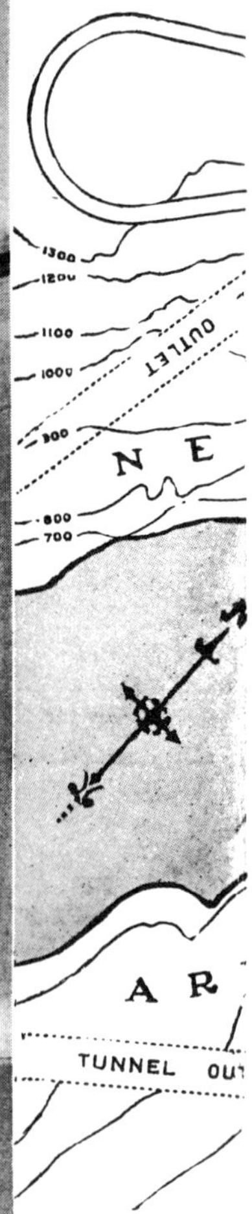


THE power that through the ages has worn the mile-deep chasm of Grand Canyon, cut black gorges through mountain ranges, and thrown a delta entirely across the Gulf of California to form the Imperial Valley, will soon be directed to more useful pursuits when the generators in the Boulder power plant commence their hum of creation and conservation.

Many visitors to Black Canyon are not particularly impressed by the size of the power-house structure as viewed from the canyon rim, for its nineteen stories of height and four acres of roof are dwarfed by the encompassing canyon walls and by the dam that rises immediately upstream. Yet more difficult is it to realize that in that relatively small space enough electricity will be produced at plant capacity to furnish each and every family in the United States with light from a 40-watt lamp, or sufficient energy to supply all the domestic light and power needed by the 8,500,000 inhabitants of the Colorado River Basin states.

The generation of electrical energy at Boulder Dam is mentioned last among the purposes of the project as set forth in the Congressional Act that authorized construction. Nevertheless, without the income from that source the building of the dam and its outlet works would have been economically infeasible, because the tillers of the lands that are benefited by flood control, silt control, and water storage could not afford to foot the bill. By the present arrangements, through contracts for power signed before work was initiated, the entire $108,800.000 for construction and an additional $11,200,000 of interest charges will be repaid the Government by 1985.

The site chosen for the U-shaped power house is immediately downstream from the dam, one wing nestling at the foot of each canyon wall, and the central part, connecting the two wings, resting on the downstream toe of the dam. Measured by chain and level rod in boresome feet, each wing is approximately 650 feet long next to the cliff and 150 feet wide at the generator-floor level, while the central section is 400 feet around at the roof and dam intersection and rises 229 feet above the lowest foundation concrete, or 154 feet above the tailrace low-water surface. Using more familiar yardsticks—the distance around the U is that of five ordinary residence blocks, while the roof parapet is nineteen stories above the lowest concrete and twelve stories above the tailrace. The roof has an area equivalent to that of two city blocks, and beneath it are about ten acres of floors.

Six Companies Inc. has built the power house from materials furnished by the

*Twenty-first of a series of articles on the Colorado River and the building of the Boulder (formerly Hoover) Dam.

†Associate Engineer, U. S. Bureau of Reclamation, Boulder Canyon Project.

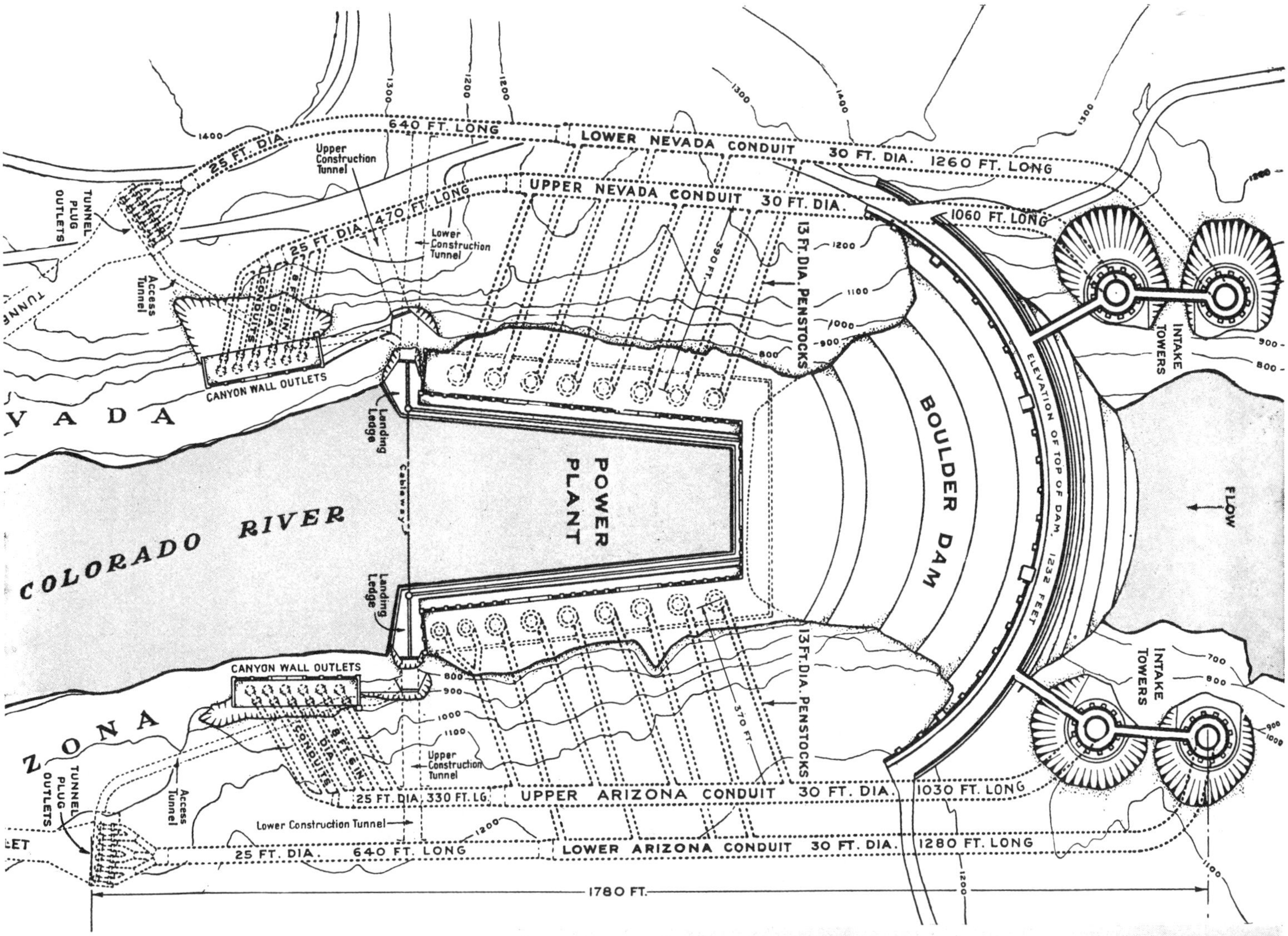


## PLAN OF OPERATION

The drawing above shows how the water which enters the intake towers will be carried through the canyon walls and either directed to the turbines of the power house or discharged into the river downstream from the dam through the outlet works. The diameters given in connection with the various tunnels refer to the steel pipes which are being installed within them by The Babcock & Wilcox Company.

At the right is a composite picture made by imposing an artist's drawing upon a photograph in order to show how Boulder Dam will appear when the reservoir is filled and the power house is in operation.

Government and has set up the cranes, but most of the 80,000,000 pounds of machinery is being placed by the forces of the U. S. Bureau of Reclamation. The larger generators and the automatic elevators in the dam and power plant are being installed by the manufacturers.

Although in its general details the structure was erected in a manner similar to city buildings, its construction was unique in that the materials were usually lowered from 600 feet above instead of being raised from ground level. All concrete was mixed at the contractor's high-level plant and transported, generally, by train and 25-ton cableway in 4-cubic-yard buckets of the spout-dump or agitator type. Compressed-air and electric vibrators were used during the pouring of all wall, column, and beam concrete. Among the materials that already have been placed in the huge structure are 240,000 cubic yards of concrete, 22,000,000 pounds of reinforcing steel, and eleven miles of pipe and conduit.

The roofing steel—with many trusses 73 feet long, 13½ feet high, and weighing 75,000 pounds—was first lowered to a landing platform at generator-floor elevation and then raised into position by the contractor's cableways, the track cable being lowered 100 feet at the center of the span so that it might safely carry the load. The structural steel in the roof has a combined weight of 11,600,000 pounds and is overlain by a composition, 4½ feet thick, consisting of alternate layers of reinforced concrete and sand and gravel topped by a waterproofing layer of bituminous-sand mastic.

The installation of power-plant machinery by Government forces is now concentrated on four 115,000-hp. units in the upstream end of the Nevada wing, on two station-service units of 3,500 hp. each in the central section—all of which will be operated by the Bureau of Power & Light of the City of Los Angeles, and on one 55,000-hp. unit in the downstream end of the Arizona wing that will be operated by the Southern Sierras Power Company. Work on the placing of two 115,000-hp. units for the Metropolitan Water District of Southern California is not expected to get underway until 1936, and the remaining ten units, making an ultimate capacity of 1,835,000-hp., are to be added within the next few years as they may be required to keep pace with the increasing demand for power. The first electricity will probably be produced in the spring of 1936.

In order more easily to visualize the operation of the plant and to understand the reason for such a large structure, let us in imagination visit the power house as it will appear when finally completed. But before starting, it might be well to follow, in a general way, the transformation of power from falling water to electrical energy and the route of the electrical circuit from the generator terminals to the transmission lines. As the several units vary primarily only in size, a description of the operation of one of the fifteen 115,000-hp. units will suffice.

WITHIN THE POWER HOUSE

**Looking downstream from within the Arizona wing, showing positions for two of the generator units. The Arizona canyon-wall outlet works may be seen under construction above.**

Water from the reservoir flows through gates in one of the four intake towers into a 30-foot-diameter pipe, the penstock header, and thence downstream to four 13-foot-diameter pipes, the power penstocks. If the operator desires, it may be continued past the penstocks to needle valves in outlet works, thus by-passing the power house. Down one of the power penstocks it goes to and through a 14-foot-diameter butterfly valve in the power plant, and thence to the spiral scroll case of the turbine. Speed vanes and wicket gates—24 of them—in the inner rim of the scroll case automatically regulate the amount of water reaching the turbine runner where the water's energy, derived from mass and velocity, is transformed into mechanical energy by rotating the horizontal runner and its vertical shaft. With flow velocity greatly reduced, the water passes peacefully out to the tailrace through the center of the turbine, the turbine draft tube, and the discharge liners.

The mechanical energy thus developed by the turbine is transmitted through an interconnecting shaft to the generator rotor which, revolving in an electromagnetic field, produces electrical energy. This electromagnetic field is created by two generators, termed exciters, mounted above the main generator and rigidly connected on the main shaft. The lower one, the main exciter, furnishes the field for the main generator, while the pilot exciter, which is compound wound and self-excited, furnishes the field for the main exciter.

For those interested in figuring "how much"—at 100 per cent efficiency, 1 cubic foot of water falling 8.81 feet in one second will develop 1 hp., or 746 watts of electricity. The maximum head on the turbines will be 590 feet, or an average of 530 feet and a minimum of 420 feet. The rate of water flow through a turbine will vary from 2,000 to 3,000 cubic feet per second.

The actual operation of a turbine, however, is not so simple as just described, for in order to maintain constant frequency the energy applied to the turbine must be regulated to suit the load on the generator. This is accomplished by a governor of the fly-ball type which controls the flow of oil

The butterfly valve at the lower end of the 13-foot penstock is closed only when it is desired to cut off all flow of water to the turbine and is not used for regulatory purposes. The valve leaf is swung to open or closed position by a rotor to which oil is supplied by a pump at 1,000 pounds per square inch pressure.

Now let us follow in a schematic manner the flow of electrical energy from the generator terminals to the transmission line. Leaving the 3 phase, 60-cycle, 82,500-kva. generator terminal at 16,500 volts, it passes through the generator-voltage oil circuit breakers to the 23,000-volt, 4,000-ampere bus structure. From there it goes to three 55,000-kva., single-phase transformers outside of the power plant where it is raised to 287,500 volts for transmission purposes. From the transformer it is carried up and out of the canyon on an overhead high-voltage circuit to a switchyard located on the Nevada side of the canyon and approximately 1,500 feet from the canyon rim. By using different combinations of high-voltage oil circuit breakers in the switchyard the current may then be directed by way of the selected transmission circuit to Los Angeles, where it is received at 275,000 volts. Each generator

TAKING FORM

Although dwarfed by the dam and canyon walls, the U-shaped power house is in reality a huge structure. Equivalent in height to a 19-story building, it has an aggregate length of more than a quarter mile. Each wing section is 150 feet wide. Within the plant will be ten acres of floor space, while the roof area will equal four city blocks. These views show the power house in course of construction.

to the servo-motor cylinders which, in turn, open or close the wicket gates and control the flow of water through the turbine. The fly-balls of the governor are driven by a synchronous motor that receives its power from a small generator mounted above the main generator and on the same shaft. A change in the speed of the generator causes a change in motor speed and, consequently, a change in the position of the wicket gates.

Should an outage or other disturbance in the electrical circuit necessitate an emergency shutdown, and thus a sudden closure of the wicket gates, the turbine pressure-regulator valve situated at the upstream side of the turbine will open and the water by-pass the turbine, escaping through the relief valve to the tailrace and thus preventing excessive pressure rise in the penstock. The movable part of the relief valve is a cylinder and needle that travels upward to open and then gradually closes, thus conserving water.

may be operated from a control cubicle placed near it or, by throwing a control transfer switch to the proper position, from the main control benchboard on the top floor of the central section of the power house.

Power for project lighting and for numerous other plant purposes is supplied by two 3,500-hp., 3,000-kva. station-service units installed at the upstream ends of the power-house wings and on the same floor as the main generators. These units differ from the others in that each is of the horizontal-shaft, impulse type equipped with two bucket wheels, two nozzles and needles, and with two shut-off gate valves of the follower ring type. Their penstocks are of 24- and 36-inch-diameter piping and are connected to the four upstream penstocks—one from each of the four header systems.

Governors operating on the same principle as those described in connection with the larger units automatically regulate the flow of water to the turbine buckets by opening or closing the needle valves in the nozzles. The valves that shut off the water for the station-service penstocks are closed or opened by hydraulic cylinders actuated by water direct from the penstocks.

The characteristics of the 3,000-kva. generators are similar to those of the larger units. Electricity produced at 2,400 volts is passed through oil circuit breakers and switchgear to several groups of transformers which serve, respectively, for lighting the power house, switchyard, dam, outlet works, and Boulder City; for operating intake-tower gates, valves in outlet works, stoney gates, and oil-handling equipment; and for operating the various power-plant auxiliaries such as butterfly-valve and governor oil pumps, generator and turbine lubricating pumps, power-house cranes, elevators, air compressors, as well as ventilating, heating, cooling, and machine-shop equipment. For the last-mentioned group of services the current is stepped down to 460 volts—additional transformers still further reducing it to 115 volts for the control of a few other unit auxiliaries.

Power for oil circuit breakers, disconnecting switches, generator relays, carbon-dioxide fire-extinguishing control valves, turbine shut-down and starting solenoids, and for emergency power-plant lighting and miscellaneous equipment is provided by 120-cell storage batteries charged by three motor-generator sets. Direct current at 125 volts is supplied by a 60-cell battery and two motor-generator sets for the valves of the station-service units, for indicator lamps on the panel boards, for the annunciator systems, and for various unit auxiliaries.

Now we are ready for our imaginary tour of inspection of the power house. For our purpose the trip will take us through only the central section and the Nevada wing as they will be in the not far distant future when all the units are installed. We shall not see the downstream end of the Arizona wing with its 55,000-hp. turbine, 40,000-kva. generator, and accessories.

We drive from Boulder City over a smooth oil-surfaced highway, and park our car near the Nevada abutment of the dam. We then walk out on the dam crest for about one third of its length to the second of the four graceful towers that rise above the downstream parapet, stopping for a few minutes *en route* to gaze appreciatively upstream at the four intake towers and at the emerald-green lake and then downstream at the power house far below with the cliff-dwellerlike structures of the canyon-wall outlet works beyond.

Inside the tower, we step through heavy bronze doors into a waiting elevator cage, drop smoothly downward from the dam crest—which is 1,232 feet above sea level—past the dam inspection galleries at elevations 1,220, 975, 875 and 780, and in less than two minutes stop 44 stories below at elevation 705. After passing through the door—this time of aluminum—into a well-lighted lobby we turn right-about and walk through a tiled and stuccoed gallery for a distance of a block and a half to the downstream face of the dam and into the central section of the power plant.

Turning to the right, we follow a corridor on to a wide balcony along the face of the dam which extends 60 feet across the generator room. The floor of this room—of beautiful terrazzo—is 32 feet below our balcony, and the ceiling, supported by massive structural-steel trusses, is 45 feet overhead. Two 300-ton cranes move easily above us; but the focal points of interest are the huge cylindrical generators with their dome lights directly opposite us. There are eight of them spaced on 62½-foot centers and extending from the river wall 53 feet into the room—their form and hum of motion the perfect embodiment of power.

The generators are air cooled, and the enclosures around them are air- as well as dust-tight, so that fire in any one of them may be immediately extinguished by the injection of carbon-dioxide gas. Speaking of air-cooling, you will find that the temperature in all the power-house rooms is pleasantly cool even though thermometers outside may register 128°F. in the shade. The wings are ventilated by fans that take air from the penstock tunnels and force it through the various galleries, while the rooms in the central section are air conditioned by individual surface coolers through which cold water is circulated.

Twenty feet beneath us, and running 620 feet to the downstream end of the generator room, is another balcony. On it and at each generator are control and excitation cubicles containing control boards for separate-unit operation. Fastened to the walls on each side of the room we note the panels of multicolored lights and the horns and bells that comprise the annunciator and signal systems which call operators to the telephone or sound an audible alarm in case of trouble in the unit operation.

One of the station-service units is at our end of the room near the canyon wall. It occupies a space 21 feet square and 6 feet high. The two impulse wheels of the turbines are mounted on a horizontal shaft and drive the 3,000-kva. generator set between them. Control boards are directly behind the unit in wall recesses.

Returning to the power-house elevator near our end of the balcony we are lowered 73 feet—six stories—and, upon leaving the cage, stop to look downstream through the long line of arches between the rooms in the pipe gallery which extends the full

NEARING COMPLETION

A picture taken last May when water storage began. By comparing it with the view on page 17, it will be seen that the structure is fast nearing its ultimate appearance.

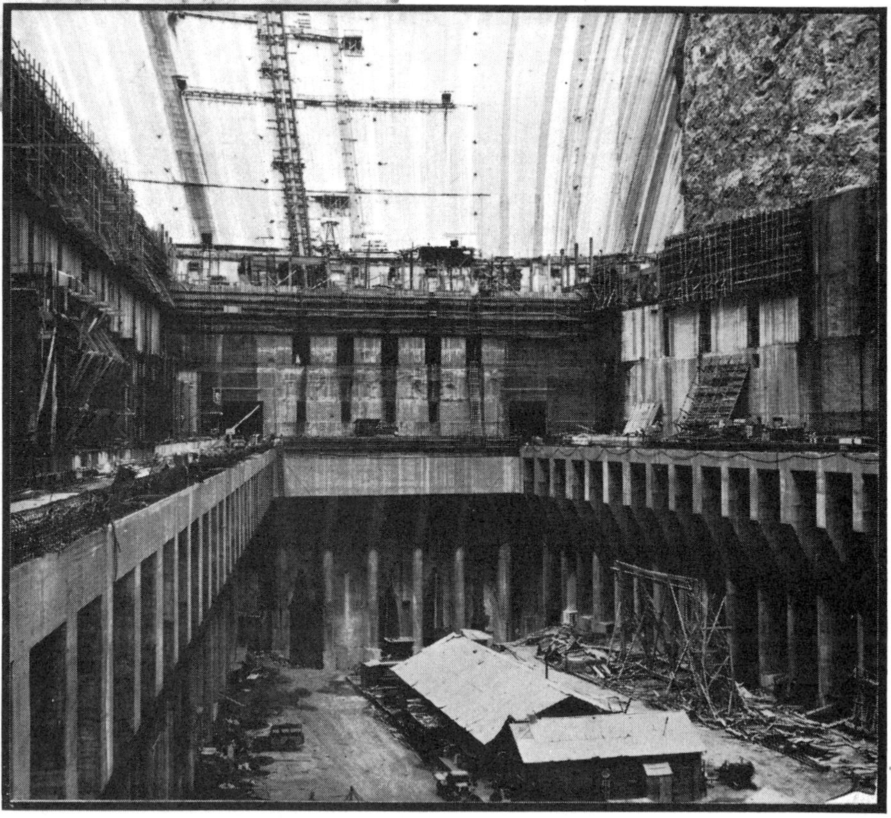

MAKING THE EQUIPMENT

**The hugeness of the generating units is brought out forcefully by the pictures on this page and on the facing one. Above is a generator-shaft assembled for inspection in the factory of the Westinghouse Electric & Manufacturing Company. At the right is shown the spiral steel casing for one of the 115,000-hp. turbines in the Allis-Chalmers plant at Milwaukee.**

length of the wing. Here are pipes and valves for miscellaneous purposes as well as pumps which are used to supply water from the tailrace for generator cooling and to evacuate water from the draft tubes in case of repairs or when it is desired to operate the generator as a synchronous condenser. To our left we see the station-service penstock running the length of the central section.

Walking through a passageway toward the cliff wall we enter the butterfly-valve gallery. There at the ends of the penstock pipes are the eight huge valves, each of which with its massive concrete base almost fills a space 15x16 feet in diameter and 20 feet from floor to ceiling. One of these valves with its operating rotor that projects into the governor gallery overhead weighs 420,000 pounds.

Retracing our steps to the elevator shaft we climb nearby stairs to elevation 645.5 and, going toward the cliff wall, we find ourselves in the governor gallery. Here we inspect one of the oil-pressure-operated rotors that move the butterfly-valve leaves, each of which we learn is cylindrical in shape, 7 feet 3 inches in diameter, and 9 feet high. A control panel is mounted on each rotor, and its motor and oil pump are installed at the base. Other equipment in the governor gallery—which, incidentally, has a 26-foot-high ceiling and is 437 feet long—are the lubrication control panels for the turbines and generators, the 460-volt control panels for the auxiliaries of each unit, the thermal board with its thermometers that register the temperature of the unit bearings, a turbine flow meter for recording the water flow, and the governor-oil pressure tanks and actuators. The latter contain the fly-ball equipment, numerous gages and valves, oil pumps and motors, and control apparatus.

Pumps for conveying dirty oil to the filter plant from governors, transformers, circuit breakers, and lubricating-oil sumps are situated at the downstream end of the gallery. At the upstream end are the circuit breakers and reactors in the line from the 16,500-volt transfer bus in the bus gallery to the 16,500/2,300-volt transformers and the 2,300-volt switch gear at elevation 717.7. This circuit will supply power to the 2,300- and the 460-volt power circuits in case the station-service units are not operating.

Walking toward the center of the wing and going down a few steps we reach the turbine gallery at elevation 643. Here are the servo-motors and shifting rings for opening and closing the turbine wicket gates and the relief valves; and rising through the gallery we see the 38-inch-diameter shafts running from turbines to generators. Embedded in the concrete beneath our feet are the turbine scroll cases—40 feet across, 10 feet in maximum outer diameter, and weighing 245,000 pounds—as well as the relief valves of spherical outer form. They are 11 feet in diameter, and their needle valves are approximately 6 feet in diameter.

Passing onward toward the river wall we enter the pipe gallery—another long room unobstructed by partitions and extending the full length of the wing. In it are pipes and more pipes, large and small, some for compressed air, others for service and cooling water, some are oil headers, and others are for Boulder City supply.

Walking toward the dam we come to the pipe shop in the central section where are situated three centrifugal pumps for Boulder City water supply, three deep-well sumps for lifting drainage water from the dam and the power house into the tailrace, and three compressors and pressure tanks which provide air for evacuating water from the turbine draft tubes, for actuating the generator air brakes, and for general use in the machine shop and elsewhere. In rooms across the face of the dam we find the 2,400- to 240/120-volt lighting transformer banks, the 2,400- to 480-volt power transformers, and four 460-volt distribution panels. Two of these connect with the unit-auxiliary boards in the wings and the other two with the boards operating miscellaneous station-service equipment.

Returning to the Nevada wing we climb again, but this time only 10 feet to the cable gallery, at elevation 653, which is another long room like the pipe gallery. As we enter we pause a moment to inspect the

250-volt d-c. and the 460-volt a-c. unit-auxiliary distribution boards which are set in recesses along the inner wall. The cabinets contain air circuit breakers for automatic switching from 460-volt a-c. to 250-volt d-c. In the event of an outage in one of the 460-volt circuits furnishing power to some unit auxiliary, a relay starts to operate, thus cutting in the 250-volt d-c. supply. In the center of the gallery are tier upon tier of racks, while to one side, hanging on steel pillars as well as along the river wall, are row upon row of conduits. On the racks are the control cables—the nervous system of the power plant. We follow them to the central section where they turn riverward up a ramp and then, near the middle of the building, turn toward the dam through a low gallery at elevation 665.2. At the dam face they abruptly disappear upward in a shaft that we later learn opens into the cable racking room at elevation 717.67.

Retracing our steps to the first stairs in the cable gallery we climb another 10 feet to the bus gallery—a duplicate in size of the one we have just left. Near the center of the room are the pairs of transformer buses and against the river wall the generator transfer bus. These buses are designed to carry 4,000 amperes, and they operate at 16,500 volts. They are made up of two 6-inch copper channels placed flange to flange, forming a 6-inch hollow square section, and each phase is mounted on porcelain insulators and inclosed in a non-magnetic metal covering. Also in this gallery, and located along the inner wall, are the generator neutral-grounding circuit breakers and reactors.

Up 10 feet more and we are on the generator floor at elevation 673. As we have already seen this wing room from the balcony at elevation 705, we shall casually inspect the completely equipped machine shop that takes up nearly all that part of the central section and then pass through the wide doorway of the shop out on to a wide platform that runs the full length of the wings and central section and connects with the cableway landing platforms at the downstream ends of the power house. There are three tracks on the wing platforms—the one nearest the river being of 12-foot gauge and the two others of 4-foot-8½-inch gauge. The former serves a 15-ton gantry crane that places bulkheads across turbine draft tubes when they are to be unwatered, and on it also travels a 180-ton car used for transferring transformers and oil circuit breakers.

The power transformers and the generator oil circuit breakers are erected on another deck along the power-house wall and 4 feet above where we are standing—all these pieces of equipment being mounted on flanged-wheel trucks. Switch houses completely inclose the circuit breakers so that any fire in the breakers can be quickly extinguished by means of carbon-dioxide gas.

Closer examination of one of the 55,000-kva. transformers reveals what an enormous piece of machinery it really is. Substantially taking up the 24x22x32-foot inclosure, it weighs, when filled with oil, more than a third of a million pounds. Heat is removed from the oil by water cooling coils. Oil is replaced periodically from the filter plant—the old oil being returned for treatment to that plant which is located in the upper construction adits through which the 30-foot-diameter pipes were taken into the header tunnels.

We stop a minute to admire the modernistic effect of the power-house exterior and to remark upon the cool air current that rises from the clear, sparkling water of the tailrace. We then return to the elevator in the machine shop and ride up to the two huge control rooms at elevation 743 in the central section. They occupy the middle 62x117 feet and contain the auxiliary, main-station-service, and supervisory control boards. Most of the operations of the power plant are recorded here by various meters, and the principal circuits can be controlled from one or the other of the boards. Orders may also be given from this station to all operators by the annunciator and telephone systems. Elsewhere on this floor are the automatic and carrier-current telephone rooms (the City of Los Angeles uses the transmission cables for telephone service) and the 80-cell telephone battery.

Now we walk downstairs to the floor at elevation 730.33 where we find the terminal boards for all the control boards just mentioned and, along the river wall, a group of offices, including those of the plant superintendent and watermaster. Down another flight of stairs at elevation 717.67 and in the center near the dam are the two 120-cell 250-volt storage batteries, the 60-cell 125-volt storage battery, five motor generator sets, a water still, and a battery-control switchboard. Also in the center of this floor are cables which come down on to racks from the terminal boards above and then enter the shafts through which they either drop to the cable gallery or rise to the top of the power house and pass into and through the control-circuit tunnel to the switchyard half a mile away and 1,500 feet back from the Nevada rim. The switchboard for distributing 2,300-volt current to various places throughout the power house is on the same floor in a long room next to the river wall.

One more flight down and we are back at elevation 705 where we began our tour of inspection of the power house. We now retrace our steps and take the elevator to the dam crest high above. We are probably pretty tired, too, as we have been in the plant several hours, have walked more than a mile, and have climbed up and down the stairs of what is equivalent to a 12-story building.

It is not to be expected that all the purposes and the interrelationship of equipment and control can be entirely understood after one short trip, but it is hoped that an impression has been gained of a smoothly operating unit producing a tremendous amount of power. The entire project—dam, power house, and related works—is truly a monument to the scientific skill and engineering ability of the Bureau of Reclamation, of which Commissioner Elwood Mead is administrative head; R. F. Walter, chief engineer; S. O. Harper, assistant chief engineer; L. Savage, chief designing engineer; L. N. McClellan, chief electrical engineer; and Walker R. Young, construction engineer for Boulder Dam and power plant.